Agricultural Machinery and Technologies

Agricultural Machinery and Technologies

Jack Atkinson

Larsen & Keller
www.larsen-keller.com

Agricultural Machinery and Technologies
Jack Atkinson
ISBN: 978-1-64172-441-8 (Hardback)

Larsen & Keller
Published by Larsen and Keller Education,
5 Penn Plaza,
19th Floor,
New York, NY 10001, USA

Cataloging-in-Publication Data

Agricultural machinery and technologies / Jack Atkinson.
p. cm.
Includes bibliographical references and index.
ISBN 978-1-64172-441-8
1. Agricultural machinery. 2. Agricultural innovations. 3. Farm mechanization.
4. Farm equipment. 5. Agricultural instruments. I. Atkinson, Jack.
S675 .A37 2020
631.3--dc23

Printed and bound in China.

For more information regarding Larsen and Keller Education and its products, please visit the publisher's website www.larsen-keller.com

Table of Contents

Preface

The machinery and technology primarily used in farming and other agricultural processes are referred to as agricultural machinery. Some of the equipment which fall in this category are hand tools, power tools and tractors. There are different types of farm machinery such as combine harvesters, tillage implements and planters. Machines are also used to deliver fertilizers and pesticides. Some of the diverse technologies used for agricultural purposes are computer monitoring systems, GPS locators and self-steer programs that are used in tractors that use less fuel and do not waste seeds or fertilizer. The topics included in this textbook on agricultural machinery and technologies are of utmost significance and bound to provide incredible insights to readers. While understanding the long-term perspectives of the topics, it makes an effort in highlighting their impact as a modern tool for the growth of this field. This book will provide comprehensive knowledge to the readers.

To facilitate a deeper understanding of the contents of this book a short introduction of every chapter is written below:

Chapter 1- Mechanized agriculture is the process that makes use of machinery in order to mechanize the work of agriculture which increases farm productivity. The technology used for the production of machinery used on a farm to help with farming is known as agricultural technology. This is an introductory chapter which will introduce briefly all the significant aspects of mechanized agriculture.

Chapter 2- Farm machinery is the machinery used in agricultural and farming activities. A few examples of such machinery are farm tractor, combine harvester, wind turbines, seed processing equipment, etc. The diverse applications of these machines have been thoroughly discussed in this chapter.

Chapter 3- Remote sensing is a process that involves the acquisition of information about an object or phenomenon without making physical contact with the object. It is used in horticulture, assessment of crop damage, crop identification, etc. This chapter has been carefully written to provide an easy understanding of the applications of remote sensing in agriculture.

Chapter 4- The application of controlled amounts of water to plants when needed is known as irrigation. It helps in the growth of agricultural crops, revegetating disturbed soils and maintaining landscapes in areas that are dry during periods of less rainfall. All the diverse methods of irrigation as well as the technologies related to it have been carefully analyzed in this chapter.

Chapter 5- The process of gathering ripe crops from the fields is known as harvesting. The stage which comes after harvesting is known as postharvest. It includes various processes such as cooling, cleaning, sorting and packing. There are different types of harvesting machines such as potato harvesters and forage harvesters. The topics elaborated in this chapter will help in gaining a better perspective about the various technologies involved in harvesting and postharvesting.

I owe the completion of this book to the never-ending support of my family, who supported me throughout the project.

Jack Atkinson

1

Mechanized Agriculture: An Introduction

Mechanized agriculture is the process that makes use of machinery in order to mechanize the work of agriculture which increases farm productivity. The technology used for the production of machinery used on a farm to help with farming is known as agricultural technology. This is an introductory chapter which will introduce briefly all the significant aspects of mechanized agriculture.

Mechanized agriculture is the process of using agricultural machinery to mechanize the work of agriculture, greatly increasing farm worker productivity. The effective mechanization contributes to increase in production in two major ways: firstly the timeliness of operation and secondly the quality of work. The requirement of power for certain operations like seedbed preparation, cultivation and harvesting becomes so great that the existing human and animal power appears to be inadequate. As a result, the operations are either partially done or sometimes completely neglected, resulting in low yield due to poor growth or untimely harvesting or both.

Benefits of Mechanization of Agriculture

1. It Increases Production: Mechanization increases the rapidity and speed of work with which farming operations can be performed. According to D. R. Bomford, "The ploughman with his three-horse am controlled three- horse; power, when given a medium-sized crawler tractor controlled between 20 to 30 horse power. His output, there-fore, went up in the ratio of about 8: 1."

   According to B. K. S. Jain, "In the U.S.A. a labourer who formerly ploughed one acre of land with a pair of horses is now able to account for 12 acres a day with a gasoline-driven tractor. By this quickening of agricultural practices the human labour required is minimised. Over a period of three decades in U.S.A., a study revealed that one-third increase was due to the use of chemicals: another one third due to better varieties, and wealthier seeds, while another one-third was due to improved farm machinery."

   According to Roy D Laird, "A more recent and more spectacular development in mechanization of agriculture has been brought in the U.S.S.R., where four times the agricultural output became that of 1913 and grain production alone increased by 70 per cent by 1960. By 1965 Socialist Competition, increased electrification and more machinery were supposed to induce a 100% increase in the efficiency of agricultural labour in that country."

2. It Increases Efficiency and Per Man Productivity: Mechanization raises the efficiency of labour and enhances the farm production per worker. By its nature it reduces the quantum of labour needed to produce a unit of output. In the U.S.A., "the amount of human labour used to produce 100 bushels of wheat dropped from 320 hours in the year 1830 to 108 hours in 1900; by 1940 a new series of improvements has reduced labour requirements to 47 hours."

According to Hecht and Barton, "Before the World War I. It took, about 35 man hrs. to grow and harvest an acre of corn; 15.2 hrs. for an acre of wheat and 15.7 hrs. for an acre of oat. In 1945-48, the labour requirements were 23.7, 6.1 and 8.1 man hours respectively. The combined, effect of fewer hours and more bushels per acre has resulted in more than halving labour requirements per unit of production. The number of man-hours required in 1910-14 per 100 bushels of corn was 135, of wheat 106 and of oat 58; in 1945- 48, the corresponding figures were 67, 34 and 23 respectively."

"It is estimated that productivity per man on farms in U.S.A. is about four and a half times that in the U.S.S.R." "In the U.S.S.R. in collective farms, production has raised labour productivity to a high level compared with the pre- revolutionary days; now labour is three times more productive there."

3. Mechanization Increases the Yield of Land Per Unit of Area: S.E. Johnson holds that "of 28 per cent increase in farm output in U.S.A., above the average of 1934-39 only about one-fourth is due to better weather, probably less than 15 per cent has resulted from expansion of crop, land acreage and the rest, about 60 per cent is largely accounted for by the fuller use of the improvements in crops, live stocks and machinery. Increase in the yield of crops, due to mechanization of farms, has been traced from 40 to 50 per cent in the case of maize; 15 to 20 per cent in Bajra and Paddy; 30 to 40 per cent in Jowar, Groundnut and Wheat."

4. Mechanization Results in Lower Cost of Work: It has been accepted by all that one of the methods of reducing unit costs is to enlarge the size c the farms and go in for more intensive farming. It is found that the cost of production and the yields can be adjusted properly if mechanization is resorted to.

5. It Contracts the Demand for Work Animals for ploughing water lifting, harvesting, transport etc. In actual operation, costs amount to little when machines are idle, whereas the cost of maintenance of draught animals remains the same during both periods of working and idleness, because animals have to be fed whether they are doing work or not. It is advantageous to use tractors when a great deal of work has to be done in a short time.

6. It Brings in other Improvements in Agricultural Technique: In its training come improvements in the sphere of irrigation, land reclamation and the prevention of soil erosion. The present-day dependence on the monsoon as the only irrigation of crops in India can be obtained by a more scientific approach.

   Besides, ploughing by tractor reclaims more land and thereby extends the cultivated area as the tractor smoothens hillocks, fills in depressions and gullies and eradicate deeps-rooted weeds. It also prevents soil erosion. Besides mechanical fertilization, contour bunding and terracing are done by mechanical methods with the help of self-propelled graders and terraces.

7. It Modifies Social Structure in Rural Areas: It results in a significant modification of the social structure in rural areas. It frees the farmers from much of the laborious, tedious, hard work on the farms. The pressure on land decreases and the status of the farmers improves.

8. It Leads to Commercial Agriculture: Mechanisation results in a shift from 'subsistence farming' to 'commercial agriculture. This shift occurs mainly due to the need for more land and capital to be associated with farmer in order to reap the full technological benefits.

   This in its turn gives rise two tendencies:

   - Gradual replacement of domestic or family by commercial methods, and
   - Search for international markets for agricultural produce.

9. It Solves the Problem of Labour Shortage: In countries where human labour falls short of requirements in agriculture, use of machines can replace human and animal power.

10. It Releases Manpower for Non-Agricultural Purposes: Since the mechanisation of agriculture results in the employment of lesser number of persons on farms, surplus manpower may be available for other economic activities.

11. It Results in Better use of Land: Mechanisation also results in better utilization of agricultural land for "the substitution of gasoline tractor for animal power means reduced demand. The use of machine energy, therefore, leads to good agricultural production, to trade many crops or saleable animal products in short, to an exchange economy and a system of land utilization in which cultivator rests on a different and infinitely more complex basis than is found in the local self-sufficient economy."

12. It Increases Farm Income: With the introduction of mechanisation the farm income as well as the individual income goes up. E. G. Nourse writes, "It accounts for the unparalleled rise of national income and with it the standard of living, it builds cities, it raises an ever loftier superstructure of financial, commercial and other cultural institutions; it turns loose economic agglomerates into social economies to closely knit by a thousand lines of interdependence. It creates much of the capital surplus on which modern economic progress is largely based. It constitutes, the lion's share to the public funds which support education, health and law and order. In short, not only do machine industry, and mechanisation and science render agriculture efficient, they create the very world in which this efficient agriculture can sell its bountiful crops."

13. It Reduces Fodder Area and Enlarges Food Area: "With the introduction of mechanisation in agriculture the surplus animal power would be reduced so that large areas of land required for producing fodder for it can be utilised for producing food for human consumption. The remaining cattle population would be better attended to and better fed under mechanised agriculture, for new and nourishing varieties of feeding stuff would be grown in cultural waste lands after reclaiming them for cultivation."

Important Arguments against Mechanisation

1. Small Sized Farms: The existence of a large farm is an essential condition for mechanisation. For proper and best utilisation of agricultural machines, holdings will have to be large and should be (bund together and not scattered in tiny plots as is the case in India. In U.S.A. the average size of a holding is about 145 acres; in Canada it is 235 acres and in U.S.S.R. it is 1,600 acres.

Mechanisation has no scope in India because of the extremely small size of holdings which arc between 3 and 12 acres. Even these small holdings are not found together but scattered over the village in tiny bits. A tractor cannot be used to plough a quarter of an acre plot. This is not a valid criticism because such farm machinery like a pump set can be installed even in a small farm of half an acre.

2. Surplus Agricultural Workers: The basic defect of mechanisation is that it will result in too many agricultural workers becoming surplus. Millions of farmers will be thrown out of land and will have to be provided alternative sources of employment.

   It is impossible to provide alternative employment for millions of persons. In U.S.A. and Canada, the real problem is shortage of labour and to overcome this difficulty, machines were invented and used, as labour saving devices.

   But India has abundant labour and there is the necessity to use this labour and not keep, it idle. What is, therefore, useful and necessary for the Western countries, need not be so for India too. Use of farm machinery may create unemployment only in the short period. In the long run, there will be more employment opportunities.

3. Surplus Cattle: The adoption of farm machinery will throw not only men out of employment but it will render the existing cattle population surplus and unnecessary. To cut down the existing cattle population will be a difficult problem. But as is well known, it will indeed be beneficial for India to reduce somehow its cattle population.

4. Poor, Illiterate and Ignorant Farmers: The Indian farmers are, in general, poor and, therefore, will not be able to buy expensive tractors and other farm machines. Besides, the farmers are uneducated. They will not be able to understand the use as well as the working of expensive farm machinery. This criticism can be easily answered. Farmers can always join together and purchase expensive farm machinery. Or the village co-operative society can purchase it and hire it out to farmers.

5. Imports: India may not be in a position to produce farm machinery on a large scale. Necessarily, therefore, she will have to depend upon foreign countries. This is only a short period problem.

6. No Increase in Productivity of Land: Mechanisation may not increase productivity of land. In India, the crucial problem is to increase the productivity of land, because land is a scarce resource of the country. The increase in the productivity of land is much more important than the increase in the productivity of labour. In a country like Japan, where mechanisation of farming is not adopted, productivity per hectare has been maximised because of intensive cultivation.

7. Lack of Spare Parts and Service Facilities and Shortage of Power: There is also lack of spare parts and service facilities in the rural areas, and an acute shortage of kerosene, petroleum and diesel oil. These need to be imported from abroad at a high cost and this might lead to a heavy drain on foreign exchange reserve.

Progress of Mechanisation of Agriculture

M.L. Darling has rightly affirmed that the plough that looks like a half open pen knife over just scratch's the soil, the hand sickle made more for a child than for a man, the old fashioned winnowing fray that wodes the wind to shift the grain from the chaff and the receds Choffer with its waste of fodder are misplaced from their primitive but immemorial functions." Thus it has a deep rooted impact on the economy of the region.

Tractor is the basic mechanical input which largely determines the extent of use of allied machinery and equipment. There was rapid progress in the number of tractor in the country. In 1961 the country had 31000 tractors which increased to 2, 52,000 in 1966. This increase is indeed too phenomenal to be overlooked. This fact can be attributed to the incidence of this period with the green revolution. Consequently, the number of tractors increased to 4, 55,000 in 1990 and is expected to have further increase to over, 1, 90,990 during 2000-01.

The number of tractors increased upto 7, 53,286 in 2002-03. The number of tractors have further increased upto 2,85,00,453. Similarly the number of tube will operated with electricity increased from 1,06,000 in 1961 to 4,55,600 in 1990 and further expected to be double of during 2000-01. The number of electric tube-wells further increased to 10, 85,000 in 2000-01. The number of power tiller was accorded 16,018 year ending 2000-01. The number of tiller increased upto 46,472 in 2005-06.

There are practical difficulties in the way of introduction of the machines on the farms. Some of these can be removed:

1. The Government should provide credit facilities to those farmers who are willing to purchase the machinery individually.
2. Joint farming societies may be developed to serve as machinery cooperatives in the different States.
3. Machine Stations of the type of M.T.S. or U.S.S.R. may be developed in different parts to give the tractors and servicing facilities to the cultivators on subsidized rates.
4. Cheaper types of small machines suitable for Indian conditions should be evolved. These would help the labourer to perform his task more efficiently rather than displace him. In this connection we would do well to remember what F.A.O. Development Paper has remarked; "Mechanisation should not be introduced in a hurry, or on a too large scale. To be successful it should be gradually expanded and kept within proficiency standards of those who operate it."

Report on India's Food Crisis and Steps to meet suggests, Special studies should be made of the need for tractor drawn ploughs or other tillage implements, with a view to procurement and use:

1. Where the soil areas will yield for greater increases in food production than is possible with other tillage implements, and where the cultivators have the ability, willingness, and organisation to make effective use of the implements without significant subsidy beyond loans.
2. Where neglected and compacted soils of derelict village commons can be brought into use.
3. Where new land development requires heavy initial ploughing or earth moving. Even scarce foreign exchange should be allocated for such machines where the benefits are very substantial.

"At the outset, it might appear that the scheme of overall mechanisation is not feasible under the present agrarian structure in India, for agricultural sector may not presently invest huge sums of money; and it would be difficult to create big farms required for mechanised agriculture compulsorily."

Therefore, we suggest that the Government should extend the scheme gradually on the following lines:

1. Complete mechanization should first be extended to the state farms.
2. The vast, sub-marginal newly reclaimed areas should be brought under mechanisation.
3. It should be extended over to such lands where co-operative joint farming societies have been formed.
4. It should also be extend to the old co-operative farms which have enough areas in compact blocks and have enough scope for mechanisation of agriculture.
5. Private big farmers should be induced to adopt mechanisation, "for the use of more efficient equipment is one of the principal ways by which productivity per man and per acre, and hence living standards can be raised."

Agricultural Technology

Agricultural technology refers to technology for the production of machines used on a farm to help with farming. Agricultural machines have been designed for practically every stage of the agricultural process. They include machines for tilling the soil, planting seeds, irrigating the land, cultivating crops, protecting them from pests and weeds, harvesting, threshing grain, livestock feeding, and sorting and packaging the products. People who are trained to design agricultural machinery, equipment, and structures are known as agricultural engineers.

A German combine harvester

Agricultural technology is among the most revolutionary and impactful areas of modern technology, driven by the fundamental need for food and for feeding an ever-growing population. It has opened an era in which powered machinery does the work formerly performed by people and animals (such as oxen and horses). These machines have massively increased farm output and dramatically changed the way people are employed and produce food worldwide. A well-known example of agricultural machinery is the tractor. Currently, mechanized agriculture also involves the use of airplanes and helicopters.

Types of Machinery

Combines might have taken the harvesting job away from tractors, but tractors still do the majority of work on a modern farm. They are used to pull implements that till the ground, plant seed, or perform a number of other tasks.

Tillage implements prepare the soil for planting by loosening the soil and killing weeds or competing plants. The best-known is the plow, the ancient implement that was upgraded in 1838 by a man named John Deere. Plows are actually used less frequently in the United States today, with offset disks used instead to turn over the soil and chisels used to gain the depth needed to retain moisture.

The most common type of seeder, called a planter, spaces seeds out equally in long rows that are usually two to three feet apart. Some crops are planted by drills, which put out much more seed in rows less than a foot apart, blanketing the field with crops. Transplanters fully or partially automate the task of transplanting seedlings to the field. With the widespread use of plastic mulch, plastic mulch layers, transplanters, and seeders lay down long rows of plastic and plant through them automatically.

After planting, other implements can be used to remove weeds from between rows, or to spread fertilizer and pesticides. Hay balers can be used to tightly package grass or alfalfa into a storable form for the winter months.

Modern irrigation also relies on a great deal of machinery. A variety of engines, pumps and other specialized gear is used to provide water quickly and in high volumes to large areas of land. Similar types of equipment can be used to deliver fertilizers and pesticides.

Besides the tractor, a variety of vehicles have been adapted for use in various aspects of farming, including trucks, airplanes, and helicopters, for everything from transporting crops and making equipment mobile to aerial spraying and livestock herd management.

New Technology and the Future

The basic technology of agricultural machines has changed little through the last century. Though modern harvesters and planters may do a better job than their predecessors, the combine of today (costing about US $250,000) cuts, threshes, and separates grain in essentially the same way earlier versions had done. However, technology is changing the way that humans operate the machines, as computer monitoring systems, GPS locators, and self-steer programs allow the most advanced tractors and implements to be more precise and less wasteful in the use of fuel, seed, or fertilizer. In the foreseeable future, some agricultural machines may be made capable of driving themselves, using GPS maps and electronic sensors. Even more esoteric are the new areas of nanotechnology and genetic engineering, where submicroscopic devices and biological processes, respectively, may be used to perform agricultural tasks in unusual new ways.

Agriculture may be one of the oldest professions, but with the development and use of agricultural machinery, there has been a dramatic drop in the number of people who can be described as "farmers." Instead of every person having to work to provide food for themselves, less than two percent of the United States population today works in agriculture, yet that two percent provides considerably more food than the other 98 percent can eat. It is estimated that at the turn of the twentieth century, one farmer in the United States could feed 25 people, whereas today, that ratio is 1:130. (In a modern grain farm, a single farmer can produce cereal to feed over a thousand people). With continuing advances in agricultural machinery, the role of the farmer will become increasingly specialized.

Agricultural Engineers

Agricultural engineers work in the context of agricultural production and processing and the management of natural resources. Their specialties include power systems and machinery design; structures and environmental science; and food and bioprocess engineering. They perform tasks such as planning, supervising, and managing the building of dairy effluent schemes, irrigation, drainage, flood and water control systems. They develop ways to conserve soil and water and to improve the processing of agricultural products. In addition, they may perform environmental impact assessments and interpret research results.

Farm Equipment

Traction and Power

- Tractor
- Crawler tractor/Caterpillar tractor.

A modern John Deere 8110 Farm Tractor plowing a field using a chisel plow

A plow in action in South Africa (Soil being turned over)

Soil Cultivation

- Cultivator
- Chisel plow

- Harrow
  - Spike harrow
  - Drag harrow
  - Disk harrow
- Plow
- Power tiller
- Rotary tiller
- Rototiller
- Spading machine
- Subsoiler
- Walking tractor

Planting

- Broadcast seeder (or broadcast spreader or fertilizer spreader)
- Plastic mulch layer
- Potato planter
- Seed drill
- Air seeder
- Precision drill
- Transplanter
  - Rice transplanter

Fertilizing and Pest Control

- Fertilizer spreader
- Terragator
- Manure spreader
- Sprayer

Irrigation

- Center pivot irrigation

Harvesting/Postharvest

- Beet harvester
- Bean harvester
- Combine harvester
- Conveyor belt
- Corn harvester
- Cotton picker
- Forage harvester (or silage harvester)
- Huller
- Potato digger
- Potato harvester
- Sickle
- Swather

Hay Making

- Bale mover
- Baler
- Conditioner
- Hay rake
- Hay tedder
- Mower

Loading

- Backhoe
- Front end loader
- Skid-steer loader

Other

- Grain auger
- Feed grinder
- Grain cart
- Rock picker

Emerging Agriculture Technologies

Venture capitalists invested more than $2 billion in agriculture technology startups in 2014 and again in 2015. That trend is expected to continue in 2016 because the demand for innovative farm technology is high, and when inventors show results, modern farmers have demonstrated a willingness to embrace those inventions and new techniques. With that in mind, here are seven emerging technologies that can literally change the agricultural landscape in the years ahead.

Soil and Water Sensors

Perhaps the equipment having the most immediate effect are soil and water sensors. These sensors are durable, unobtrusive and relatively inexpensive. Even family farms are finding it affordable to distribute them throughout their land, and they provide numerous benefits. For instance, these sensors can detect moisture and nitrogen levels, and the farm can use this information to determine when to water and fertilize rather than rely on a predetermined schedule. That results in more efficient use of resources and therefore lowered costs, but it also helps the farm be more environmentally friendly by conserving water, limiting erosion and reducing fertilizer levels in local rivers and lakes.

Weather Tracking

Although we still make jokes about our local meteorologists, the truth is that computerized weather modeling is becoming increasingly sophisticated. There are online weather services that focus exclusively on agriculture, and farmers can access these services on dedicated onboard and handheld farm technology but also via mobile apps that run on just about any consumer smartphone. This technology can give farmers enough advanced notice of frost, hail and other weather that they can take precautions to protect the crops or at least mitigate losses to a significant degree.

Satellite Imaging

As remote satellite imaging has become more sophisticated, it's allowed for real-time crop imagery. This isn't just bird's-eye-view snapshots but images in resolutions of 5-meter-pixels and even greater. Crop imagery lets a farmer examine crops as if he or she were standing there without actually standing there. Even reviewing images on a weekly basis can save a farm a considerable amount of time and money. Additionally, this technology can be integrated with crop, soil and water sensors so that the farmers can receive notifications along with appropriate satellite images when danger thresholds are met.

Pervasive Automation

Pervasive automation is a buzz term in the agriculture technology industry, and it can refer to any technology that reduces operator workload. Examples include autonomous vehicles controlled by robotics or remotely through terminals and hyper precision, such as RTK navigation systems that make seeding and fertilization routes as optimal as possible. Most farming equipment already adopts the ISOBUS standard, and that puts on the precipice of a farming reality where balers, combines, tractors and other farming equipment communicate and even operate in a plug-and-play manner.

Minichromosomal Technology

Perhaps one of the most exciting advents in agriculture technology is coming in a very tiny package. A minichromosome is a small structure within a cell that includes very little genetic material but can, in layman's terms, hold a lot of information. Using minichromosomes, agricultural geneticists can add dozens and perhaps even hundreds of traits to a plant. These traits can be quite complex, such as drought tolerance and nitrogen use. However, what is most intriguing about minichromosomal technology is that a plant's original chromosomes are not altered in any way. That results in faster regulatory approval and wider, faster acceptance from consumers.

RFID Technology

The soil and water sensors mentioned earlier have set a foundation for traceability. The industry has only begun to realize this infrastructure, but it's taking shape quickly. These sensors provide information that can be associated with farming yields. It may seem like science fiction, but we're living in a world where a bag of potatoes can have a barcode that you can scan with your smartphone in order to access information about the soil that yielded them. A future where farms can market themselves and have loyal consumers track their yields for purchase is not far-fetched.

Vertical Farming

Vertical farming has been a science fiction topic as far back as the 1950s and perhaps further, and now it's not only scientifically viable but will be financially viable within the decade. Vertical farming a component of urban agriculture is the practice of producing food in vertically stacked layers. This offers many advantages. Perhaps the most obvious is the ability to grow within urban environments and thus have fresher foods available faster and at lower costs. However, vertical farming won't be limited to just urban environments like initially expected. Farmers in all areas can use it to make better use of available land and to grow crops that wouldn't normally be viable in those locations.

Artificial Intelligence in Agriculture

Agriculture is seeing rapid adoption of Artificial Intelligence (AI) and Machine Learning (ML) both in terms of agricultural products and in-field farming techniques. Cognitive computing in particular, is all set to become the most disruptive technology in agriculture services as it can understand, learn, and respond to different situations to increase efficiency.

Providing some of these solutions as a service like chatbot or other conversational platform to all the farmers will help them keep pace with technological advancements as well as apply the same in their daily farming to reap the benefits of this service.

Currently, Microsoft is working with 175 farmers in Andhra Pradesh, India to provide advisory services for sowing, land, fertilizer and so on. This initiative has already resulted in 30% higher yield per hectare on an average compared to last year.

Given below are top five areas where the use of cognitive solutions can benefit agriculture:

Growth Driven by IOT

Huge volumes of data get generated every day in both structured and unstructured format. These relate to data on historical weather pattern, soil reports, new research, rainfall, pest infestation, images from Drones and cameras and so on. Cognitive IOT solutions can sense all this data and provide strong insights to improve yield.

Proximity Sensing and Remote Sensing are two technologies which are primarily used for intelligent data fusion. One use case of this high-resolution data is Soil Testing. While remote sensing requires sensors to be built into airborne or satellite systems, proximity sensing requires sensors in contact with soil or at a very close range. This helps in soil characterization based on the soil below the surface in a particular place. Hardware solutions like Rowbot (pertaining to corns) are already pairing data-collecting software with robotics to prepare the best fertilizer for growing f corns in addition to other activities to maximize output.

Image-based Insight Generation

Precision farming is one of the most discussed areas in farming today. Drone-based images can help in in-depth field analysis, crop monitoring, scanning of fields and so on. Computer vision technology, IOT and drone data can be combined to ensure rapid actions by farmers. Feeds from drone image data can generate alerts in real time to accelerate precision farming.

Companies like Aerialtronics have implemented IBM Watson IoT Platform and the Visual Recognition APIs in commercial drones for image analysis in real time. Given below are some areas where computer vision technology can be put to use:

- Disease detection: Preprocessing of image ensure the leaf images are segmented into areas like background, non-diseased part and diseased part. The diseased part is then cropped and send to remote labs for further diagnosis. It also helps in pest identification, nutrient deficiency recognition and more.

- Crop readiness identification: Images of different crops under white/UV-A light are captured to determine how ripe the green fruits are. Farmers can create different levels of readiness based on the crop/fruit category and add them into separate stacks before sending them to the market.
- Field management: Using high-definition images from airborne systems (drone or copters), real-time estimates can be made during cultivation period by creating a field map and identifying areas where crops require water, fertilizer or pesticides. This helps in resource optimization to a huge extent.

Identification of Optimal Mix for Agronomic Products

Based on multiple parameters like soil condition, weather forecast, type of seeds, infestation in a certain area and so on, cognitive solutions make recommendations to farmers on the best choice of crops and hybrid seeds. The recommendation can be further personalized based on the farm's requirement, local conditions, and data about successful farming in the past. External factors like marketplace trends, prices or consumer needs may also be factored into enable farmers take a well-informed decision.

Health Monitoring of Crops

Remote sensing techniques along with hyper spectral imaging and 3d laser scanning are essential to build crop metrics across thousands of acres. It has the potential to bring in a revolutionary change in terms of how farmlands are monitored by farmers both from time and effort perspective. This technology will also be used to monitor crops along their entire lifecycle including report generation in case of anomalies.

Automation Techniques in Irrigation and Enabling Farmers

In terms of human intensive processes in farming, irrigation is one such process. Machines trained on historical weather pattern, soil quality and kind of crops to be grown, can automate irrigation and increase overall yield. With close to 70% of the world's fresh water being used in irrigation, automation can help farmers better manage their water problems.

Importance of Drone

As per a recent PWC Study, the total addressable market for Drone-based solutions across the globe is $127.3 billion and for agriculture it is at $32.4 billion.

Drone-based solutions in agriculture have a lot of significance in terms of managing adverse weather conditions, productivity gains, precision farming and yield management.

- Before the crop cycle, drone can be used to produce a 3-D field map of detailed terrain, drainage, soil viability and irrigation. Nitrogen-level management can also be done by drone solutions.

- Aerial spraying of pods with seeds and plant nutrients into the soil provides necessary supplements for plants. Apart from that, Drones can be programmed to spray liquids by modulating distance from the ground depending on the terrain.
- Crop Monitoring and Health assessment remains one of the most significant areas in agriculture to provide drone-based solutions in collaboration with Artificial Intelligence and computer vision technology. High-resolution cameras indrones collect precision field images which can be passed through convolution neural network to identify areas with weeds, which crops need water, plant stress level in midgrowth stage. In terms of infected plants, by scanning crops in both RGB and near-infra red light, it is possible to generate multi-spectral images using drone devices. With this, it is possible to specify which plants have been infected including their location in a vast field to apply remedies, instantly. The multi spectral images combinehyper spectral images with 3D scanning techniques to define the spatial information system that is used for acres of land. The temporal component provides the guidance for the entire lifecycle of the plant.

Precision Farming

The phrase "Right Place, Right Time, Right Product" sums up precision farming. This is a more accurate and controlled technique that replaces the repetitive and labor-intensive part of farming. It also provides guidance about crop rotation, optimum planting and harvesting time, water management, nutrient management, pest attacks and so on.

Key technologies that enable precision farming are give below:

- High precision positioning system.
- Automated steering system.
- Geo mapping.
- Sensor and remote sensing.
- Integrated electronic communication.
- Variable rate technology.

Goals for Precision Farming

- Profitability: Identifying crops and market strategically as well as predicting ROI based on cost and margin.
- Efficiency: By investing in precision algorithm, better, faster and cheaper farming opportunities can be utilized. This enables overall accuracy and efficient use of resource.
- Sustainability: Improved social, environmental and economic performance ensures incremental improvements each season for all the performance indicators.

Examples of Precision Farming Management

- Identification of stress level in a plant is obtained from high-resolution images and multiple sensor data on plants. This large set of data from multiple sources needs to be used as an input for Machine Learning to enable data fusion and feature identification for stress recognition.

- Machine learning models trained on plant images can be used to recognize stress levels in plants. The entire approach can be classified into four stages of identification, classification, quantification and prediction to make better decisions.

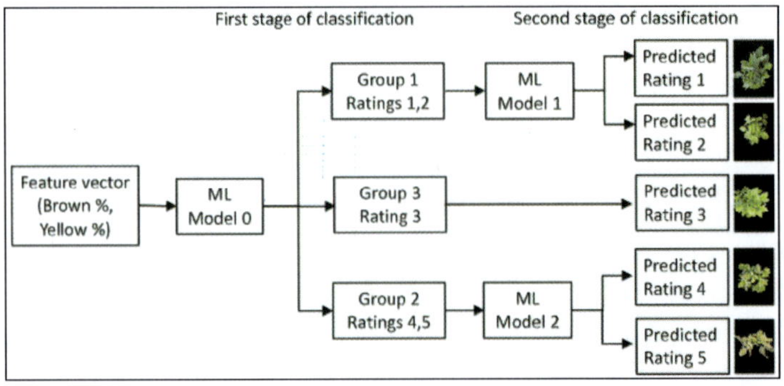

Yield Management using AI

The emergence of new age technologies like Artificial Intelligence (AI), Cloud Machine Learning, Satellite Imagery and advanced analytics are creating an ecosystem for smart farming. Fusion of all this technology is enabling farmers achieve higher average yield and better price control.

Microsoft is currently working with farmers from Andhra Pradesh to provide advisory services using Cortana Intelligence Suite including Machine Learning and Power BI. The pilot project uses an AI sowing app to recommend sowing date, land preparation, soil test-based fertilization, farm yard manure application, seed treatment, optimum sowing depth and more to farmers which has resulted in 30% increase in average crop yield per hectare. Technology can also be used to identify optimal sowing period, historic climate data, real time Moisture Adequacy Data (MAI) from daily rainfall and soil moisture to build predictability and provide inputs to farmers on ideal sowing time. To identify potential pest attacks, Microsoft in collaboration with United Phosphorus Limited is building a Pest Risk Prediction API that leverages AI and machine learning to indicate in advance, the risk of pest attack. Based on the weather condition and crop growth stage, pest attacks are predicted as High, Medium or Low.

Robotics Helping in Digital Farming

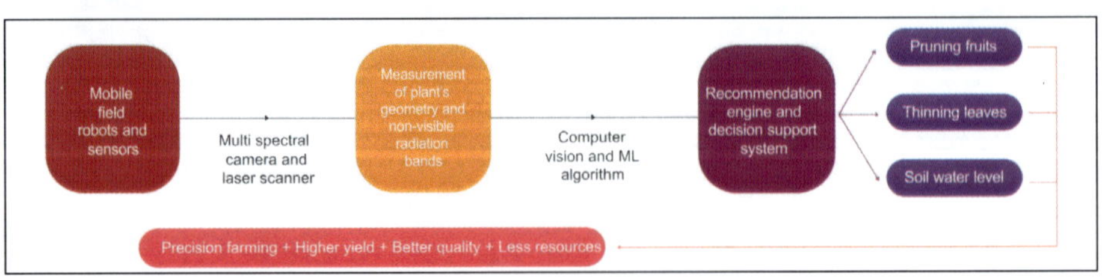

Challenges in AI Adoption in Agriculture

Though Artificial Intelligence offers vast opportunities for application in agriculture, there still exists a lack of familiarity with high tech machine learning solutions in farms across most parts of the world. Exposure of farming to external factors like weather conditions, soil conditions and presence of pests is quite a lot. So what might look like a good solution while planning during the start of harvesting, may not be an optimal one because of changes in external parameters.

AI systems also need a lot of data to train machines and to make precise predictions. In case of vast agricultural land, though spatial data can be gathered easily, temporal data is hard to get. For example, most of the crop-specific data can be obtained only once in a year when the crops are growing. Since the data infrastructure takes time to mature, it requires a significant amount of time to build a robust machine learning model.

This is one reason why AI sees a lot of use in agronomic products such as seeds, fertilizer, pesticides and so on rather than in-field precision solutions.

Applications of AI in Agriculture

The most popular applications of AI in agriculture appear to fall into three major categories:

- Agricultural Robots – Companies are developing and programming autonomous robots to handle essential agricultural tasks such as harvesting crops at a higher volume and faster pace than human laborers.
- Crop and Soil Monitoring – Companies are leveraging computer vision and deep-learning algorithms to process data captured by drones and/or software-based technology to monitor crop and soil health.
- Predictive Analytics – Machine learning models are being developed to track and predict various environmental impacts on crop yield such as weather changes.

Machine Learning in Agriculture

Machine learning is everywhere throughout the whole growing and harvesting cycle. It begins with a seed being planted in the soil — from the soil preparation, seeds breeding and water feed measurement — and it ends when robots pick up the harvest determining the ripeness with the help of computer vision.

This is how agriculture can benefit from Machine Learning at every stage:

Species Management

Species Breeding

Our favorite, this application is so logical and yet so unexpected, because mostly you read about harvest prediction or ambient conditions management at later stages.

Species selection is a tedious process of searching for specific genes that determine the effectiveness of water and nutrients use, adaptation to climate change, disease resistance, as well as nutrients content or a better taste. Machine learning, in particular, deep learning algorithms, take decades of field data to analyze crops performance in various climates and new characteristics developed in the process. Based on this data they can build a probability model that would predict which genes will most likely contribute a beneficial trait to a plant.

Species Recognition

While the traditional human approach for plant classification would be to compare color and shape of leaves, machine learning can provide more accurate and faster results analyzing the leaf vein morphology which carries more information about the leaf properties.

Field Conditions Management

Soil Management

For specialists involved in agriculture, soil is a heterogeneous natural resource, with complex processes and vague mechanisms. Its temperature alone can give insights into the climate change effects on the regional yield. Machine learning algorithms study evaporation processes, soil moisture and temperature to understand the dynamics of ecosystems and the impingement in agriculture.

Water Management

Water management in agriculture impacts hydrological, climatological, and agronomical balance. So far, the most developed ML-based applications are connected with estimation of daily, weekly, or monthly evapotranspiration allowing for a more effective use of irrigation systems and prediction of daily dew point temperature, which helps identify expected weather phenomena and estimate evapotranspiration and evaporation.

Crop Management

Yield Prediction

Yield prediction is one of the most important and popular topics in precision agriculture as it defines yield mapping and estimation, matching of crop supply with demand, and crop management. State-of the-art approaches have gone far beyond simple prediction based on the historical data, but incorporate computer vision technologies to provide data on the go and comprehensive multidimensional analysis of crops, weather, and economic conditions to make the most of the yield for farmers and population.

Crop Quality

The accurate detection and classification of crop quality characteristics can increase product price and reduce waste. In comparison with the human experts, machines can make use of seemingly meaningless data and interconnections to reveal new qualities playing role in the overall quality of the crops and to detect them.

Disease Detection

Both in open-air and greenhouse conditions, the most widely used practice in pest and disease control is to uniformly spray pesticides over the cropping area. To be effective, this approach requires significant amounts of pesticides which results in a high financial and significant environmental cost. ML is used as a part of the general precision agriculture management, where agro-chemicals input is targeted in terms of time, place and affected plants.

Weed Detection

Apart from diseases, weeds are the most important threats to crop production. The biggest problem in weeds fighting is that they are difficult to detect and discriminate from crops. Computer vision and ML algorithms can improve detection and discrimination of weeds at low cost and with no environmental issues and side effects. In future, these technologies will drive robots that will destroy weeds, minimizing the need for herbicides.

Livestock Management

Livestock Production

Similar to crop management, machine learning provides accurate prediction and estimation of farming parameters to optimize the economic efficiency of livestock production systems, such as cattle and eggs production. For example, weight predicting systems can estimate the future weights 150 days prior to the slaughter day, allowing farmers to modify diets and conditions respectively.

Animal Welfare

In present-day setting, the livestock is increasingly treated not just as food containers, but as animals who can be unhappy and exhausted of their life at a farm. Animals behavior classifiers can connect their chewing signals to the need in diet changes and by their movement patterns, including standing, moving, feeding, and drinking, they can tell the amount of stress the animal is exposed to and predict its susceptibility to diseases, weight gain and production.

This is an application that can be called a bonus: imagine a farmer sitting late at night and trying to figure out the next steps in management of his crops. Whether he could sell more now to a local producer or head to a regional fair. He needs someone to talk through the various options to take a final decision. To help him, companies are now working on development specialized chatbots that would be able to converse with farmers and provide them with valuable facts and analytics. Farmers' chatbots are expected to be even smarter than consumer-oriented Alexa and similar helpers, since they would be able not only to give figures, but analyze them and consult farmers on tough matters.

Vision Systems used in Agriculture

When it comes to machine vision, industrial automation is by far the most common application. Still, it's important to remember that agriculture has benefited from advances in technology since the invention of the plow. With every change in irrigation and cultivation techniques, it becomes possible to raise healthier crops with larger, more nourishing yields. Now, machine vision is being applied to agriculture in increasingly effective and novel ways.

The main barrier to implementing machine vision in agriculture was the need to ruggedize the key systems. To adapt machine vision technologies to rough outdoor environments, engineers needed to develop hardware and software that would maintain calibration during heavy duty use. Likewise, it was necessary to design sensor systems that would collect the information most valuable to agriculture professionals. With those challenges met, machine vision has a great future ahead of it in agriculture.

Some exciting innovations include:

1. Automation of Fundamental Processes: As in industrial environments, one of the most crucial benefits of machine vision in agriculture is its capacity to automate time-consuming, labor-intensive tasks. With further refinement of sensor systems and actuators, machine

vision systems will be well-equipped to manage fruit picking, crop control, harvesting, and a range of other tasks. Human oversight would be used mainly in a supervisory capacity to help improve the systems and their results.

2. X-Ray Detection for Weeding Systems: Each crop has its own needs. The processes and business environment around its cultivation must reflect nature's imperatives. For example, studies have shown that weed control accounts for a relatively high percentage of the overhead costs of tomato production in some areas of California. An automated weeding system developed by researchers at UC Davis improved conditions by using X-rays to detect and eliminate weeds growing near tomato stems.
3. Agricultural Logistics: Not all machine vision applications in agriculture take place "out in the field." The industry has a wealth of lessons to apply from other industries. With only minor adjustments, 2D machine vision systems that are already in use can be modified to ensure repeatable results in harvesting. Fill level inspections, part measurement, and automated counting systems are all well within the capacity of today's systems, even when the individual products are irregular.
4. Heavy Agricultural Equipment: Machine vision can enhance existing agricultural equipment, such as harvesters, and can be used to broadly automate some heavy machinery. The adoption of vision technology is helping to create a completely new generation of agricultural equipment, such as lettuce thinners. These large vehicles are able to distinguish between healthy lettuce seedlings and invasive plants and apply the appropriate chemical or manual countermeasure to improve seedling viability.
5. Drones: Drones can be used to monitor macro-level crop conditions. This gives professionals the opportunity to take action quickly against unexpected outbreaks of disease, pests, or adverse environmental conditions.

Agriculture Robots

An agbot, also called an agribot, is an autonomous robot used in farming to help improve efficiency and reduce reliance on manual labor. Future farms are expected to be tilled, sown, tended and harvested solely by fleets of co-operating autonomous robots called swarm robots that will weed, fertilize, control pests and diseases, all the while collecting valuable data.

Equipped with specialized arms, end effectors and other tools to perform a variety of agricultural tasks, agbots can connect to wireless sensor networks (WSNs) and with the help of drones, gather large amounts of data. Big data analytics will help farmers extract information from the vast amount of data to make farming more efficient and improve output.

The trend toward using agbots is expected to increase exponentially to help alleviate a shortage of human farm labor. Today, robots for fruit picking, milking and sheep shearing have successfully done jobs that previously required human manual labor. Thousands of robotic milking parlors are in place worldwide, and mobile bots are helping dairy farmers automate tasks such as feed pushing and cleaning manure.

The use of robotics at dairy farms accounts for a substantial share of the market for robots and drones in precision agriculture, which IDTechEx researchers predict will grow from $3 billion in 2016 to $10 billion as early as 2022. In addition to being unmanned and energy-efficient, IDTechEx predicts agricultural robots are apt to be slow (so that they can devote more attention to each plant), lightweight (to reduce soil compaction), and small (to keep costs down).

Agbots will be used as:

- Equipment manufacturers are developing tractors that use machine vision and autonomous robotics to replace traditional large, expensive and inefficient tractors.
- In the rules-oriented organic farming field, robotic weeding implements have been used to inspect crop rows and identify weeds rapidly.
- Robots operating at low speeds can detect weeds. Next-generation computer vision algorithms will be used to classify them.
- Robots are being used for harvesting and sorting, packaging and boxing.
- Other future applications include planting seeds, grafting plants, seeding clouds, analyzing soil and monitoring the environment.

Agricultural robots are increasing production yields for farmers in various ways. From drones to autonomous tractors to robotic arms, the technology is being deployed in creative and innovative applications.

Agricultural Robot Applications

Agricultural robots automate slow, repetitive and dull tasks for farmers, allowing them to focus more on improving overall production yields. Some of the most common robots in agriculture are used for:

- Harvesting and picking
- Weed control
- Autonomous mowing, pruning, seeding, spraying and thinning
- Phenotyping
- Sorting and packing
- Utility platforms

Harvesting and picking is one of the most popular robotic applications in agriculture due to the accuracy and speed that robots can achieve to improve the size of yields and reduce waste from crops being left in the field.

These applications can be difficult to automate, however. For example, a robotic system designed to pick sweet peppers encounters many obstacles. Vision systems have to determine the location and ripeness of the pepper in harsh conditions, including the presence of dust, varying light intensity, temperature swings and movement created by the wind.

But it still takes more than advanced vision systems to pick a pepper. A robotic arm has to navigate environments with just as many obstacles to delicately grasp and place a pepper. This process is very different from picking and placing a metal part on an assembly line. The agricultural robotic arm must be flexible in a dynamic environment and accurate enough not to damage the peppers as they're being picked.

Harvesting and picking robots are becoming very popular among farmers, but there are dozens of other innovative ways the agricultural industry is deploying robotic automation to improve their production yields.

The demand for food is outpacing available farmland and it's up to farmers to close this gap. Agricultural robots are helping them do just that.

Outdoor Agriculture Robotics

Drones/Aerial Imaging

Drones are one of the first robotic applications that farmers have adopted thanks to obvious applications like time and labor savings in not having to go out to visual checking on a crop.

These drones are equipped with multi-spectral and photo cameras that can monitor crop stress, plant growth and predict yields with more advanced drones able to carry and deliver payloads like herbicides, fertilizer, and water.

The drones can be utilized via a drone-as-a-service type operation with scheduled flyovers or can be stored on site and used as needed with weatherproof docking stations that allow the drones to recharge and send data back to be analyzed.

Spraying and Weeding Robots

Utilizing databases of weed images, these companies train their robots to detect and pluck weeds or apply pesticides directly on the weed itself and not the plant.

With a growing number of weeds become resistant to pesticides and the overall cost and use (estimated and 3 billion pounds of herbicides applied at a cost of $25B each year) this technology could be a huge boost to both the farmers bottom line and the environment.

Auto Steering and Autonomous Navigation

Using GPS farmers can now enable their tractors and other farm equipment to automatically run in your fields or at the very least reduce fatigue while you are in the cab.

Most of these devices offer a high degree of accuracy (down to a few inches) and will automatically adjust to hilly terrain and improve over time if it picks up on any alignment issues, etc. Follow your tractors exact location and progress using your smartphone.

Current system rely on operators to map out an obstacles but this is quickly changing with machine learning and the machines will automatically avoid anything unusual.

Fruit Harvesting Robotics

Pick your crop 24/7 with robotic arms. Crops that have an early focus on robotic harvesting include strawberries, cucumbers and orchard fruit like apples.

In the US in 2016 alone over 7 million tons of apples were picked by hand. Hourly wages continue to increase for this difficult job and labor shortages are a real problem. These innovation machines use a combination of image processing and robotic arms to determine what to pick. Quality control and grading of the fruit can be done in one operation keeping repeated handling down. Data analytics on the crop can be helpful in determining business revenue and organizing packaging and processing operations.

Indoor Agriculture Robotics

Greenhouse Harvesting

Using machine vision tools and robotics new harvesting systems are starting to emerge for high value crops like tomatoes and strawberries. The harvesting robots are able to go down greenhouse aisles accurately identify ripe versus unripe plants, harvest them and place them in on-board boxing systems.

Greenhouse Material Handing

With robots working safely alongside people robots can perform dreaded manual labor tasks like plant spacing.

The robots can also increase plant quality by optimizing plant placement and reducing non-labor production costs including a reduction in the amount of water, pesticides, herbicides and fertilizers used as a result of more consistent spacing.

The world population is expected to hit a whopping number of 9 billion by 2050. What is expected to follow is a dramatic rise in agricultural production, doubling to meet the coming demand. This need has caused farmers to turn to robotics as a solution for the coming future.

Now you have probably heard of how people across various industries are using robotics to disrupt their respective industries. Customer service, packaging and shipping, manufacturing, and transportation are all industries soon to be hiring more robot employees.

Nevertheless, the growing population, rise of AI and new developments in robotics has caused the world of agriculture robotics to explode with innovation.

From nursery planting to shepherding and herding, here are some of the robots already in agriculture.

Ecorobotix

Powered by the sun, this lightweight GPS, fully autonomous drone has the ability to use its solar power to run all day. The robot uses its complex camera system to target and spray weeds.

Because of its very precise arms, the robot uses 90% less herbicide, making it 30% cheaper than traditional treatments. A fleet of these robots could easily replace human farm labor down the road.

Naio Technologies

Naio Technologies have a host of robots that not only act as the perfect farm hand using techniques that preserve and protect the local environment.

The robots have the ability to weed, hoe, and assist during harvesting. As stated by the team, "We want to provide all participants in the agriculture process with access to the latest technology, to help grow healthier, more abundant and environmentally friendly crops."

Energid Citrus Picking System

Perfect for those in the citrus fruit business, the Energid systems are fast and efficient harvesting systems. The systems can pick a fruit every 2 to 3 seconds. Even more so, the robot is cheap to build, making it significantly cheaper than human labor.

Agrobot E-series

If you are expecting to have any strawberry fields in the near future, you may want to look into getting an Agrobot E-Series.

With its twenty-four robotic arms working wirelessly and an advanced AI system, the E-series cannot only pick strawberries really fast but it can identify the ripeness of a strawberry in the field.

Blue River LettuceBot2

Mentioned in the name, these robots are the perfect tools for farmers and their lettuce crops. With its imaging system, the LettuceBot2 is a popular tool in the agriculture world that attaches itself to a tractor to thin out lettuce fields as well as prevent herbicide-resistant weeds.

Again, because of this robot's ability to be precise, it uses 90% less herbicide on crops.

Agribotix

Drones will play a huge role in monitoring large areas of crops. Agribotix is a low-cost tool for farmers, to collect crop data over time, or in real-time.

From taking precise aerial photos to recording video, the company's collection of drones even has infrared sensors that can measure the health of crops while in the air.

Vision Robotics

Dubbed as some of the best products on the market, the San Diego company has been working on agriculture robotic systems for seven years.

The VR team has used AI-powered robots that can tackle a host of products include a vineyard pruner that images vines and uses a robotic arm to thin plants, as well as an automated lettuce thinner.

RoBoPlant

This robot has both semi and fully automatic machinery for greenhouse management or protected horticulture. The artificially intelligent robot is able to take flats of peat seedlings to separate them and plant them in optimal patterns.

Agricultural Drone

Drone technology is a phenomenal innovation that continues to have far-reaching effects across today's society, transforming our lives and the way we do business.

The agricultural industry seems to have embraced drone technology with open arms, using these advanced tools to transform modern farming.

High-tech drones allow farmers, and the drone pilots that operate them, to increase efficiency in certain aspects of the farming process. From crop monitoring to planting, livestock management, crop spraying, irrigation mapping, and more.

Below we dive deep into the agricultural drone market and its many applications, along with a look at the future of this exciting industry and what you can do to get started as an agricultural drone pilot.

Agricultural drones help to achieve and improve what's known as precision agriculture:

- This approach to farming management is based on observing, measuring, and taking action based on real-time crop and livestock data. It erases the need for guesswork in modern farming and instead gives farmers the ability to maximize their yields and run more efficient organizations, all while enhancing crop production.
- In recent years the cost of agriculture drones has rapidly declined, which has not only led to the explosion of drone use cases in agriculture but has made it a no-brainer investment for modern farmers.
- In fact, the agricultural drone market is expected to grow over 38% in coming years. Driven by growing population levels and changing climate patterns, the need for efficient agriculture is only going to become more important.

There are multiple uses for agricultural drones, including:

- Scouting land and crops,
- Checking for weeds and spot treating plants,
- Monitoring overall crop health,
- Managing livestock and monitoring for health issues.

Drones are equipped with technology like propulsion systems, infrared cameras, GPS and navigation systems, programmable controllers, and automated flight planning. Plus, with custom-made data processing software any collected information can instantly be put to use towards better management decisions.

Drones are transforming how agriculture and farming are done:

- By implementing drone technology, farms and agriculture businesses can improve crop yields, save time, and make land management decisions that'll improve long-term success.
- Farmer's today have a variety of complex factors that influence the success of their farms. From water access to changing climate, wind, soil quality, the presence of weeds and insects, variable growing seasons, and more.
- As a result, farmers are turning to high-level drone technology to help remedy these problems, and provide fast and efficient solutions.
- Agricultural drones allow farmers to obtain access to a wealth of data they can use to make better management decisions, improve crop yields, and increase overall profitability.
- Drones can be used to collect data related to crop yields, livestock health, soil quality, nutrient measurements, weather and rainfall results, and more. This data can then be used to get a more accurate map of any existing issues, as well as create solutions based upon extremely reliable data.
- The agriculture industry is no stranger to embracing changing technological trends to streamline business. The use of drones in agriculture is the next technological wave that'll help agricultural businesses meet the changing and growing demands of the future.

Uses of Agricultural Drones

Drone technology can help to accomplish once time-consuming and difficult tasks, all while reducing costs across the board.

You can expect the current uses of drones in agriculture to continue to evolve as the industry matures and new technology is introduced.

Currently, there are six common uses of agricultural drones, which we profile below:

Soil and Field Analysis

At the beginning, middle, and end of a crop cycle drones can be used to help obtain useful data surrounding the quality of the existing soil. By obtaining 3D maps of existing soil, you'll be able to see if there are any issues surrounding soil quality, nutrient management, or soil dead zones.

This information can help farmers determine the most effective patterns for planting, managing crops, soil, and more. Ongoing monitoring can help to better utilize water resources, and more effectively manage crop nutrient levels.

Seed Planting

Drone planting is a relatively newer technology and not as widely used, but some companies are experimenting with drone planting. Essentially, manufacturers are experimenting with custom systems that have the ability to shoot seed pods into prepared soil.

Drone startup companies have been instrumental in developing unique drone technologies to assist with a wide range of ecological and agricultural issues. For example, a company is using unmanned aircraft capable of delivering up to 57 pounds of payload in the form of tree seeds, herbicides, fertilizer and water per aircraft per flight to assist reforestation and replanting projects.

This technology helps to minimize the need for on-the-ground planting, which can be costly, time-intensive, and strenuous work. This same drone technology can be adapted and applied to a wide range of farm types, reducing overall planting times and labor costs across the board.

Crop Spraying and Spot Spraying

Crops require consistent fertilization and spraying in order to maintain high yields. Traditionally this was done manually, with vehicles, or even via airplane. These methods are not only inefficient, and burdensome, but they can be very costly as well.

With approval from the FAA, Drones can be equipped with large reservoirs, which can be filled with fertilizers, herbicides, or pesticides. Using drones for crop spraying is much safer and cost-effective. Drones can even be operated completely autonomously and programmed to run on specific schedules and routes.

For example, if there's a fungus breakout in a certain section of the crops, drones can be used to spot treat the issue. With the speed at which drones can operate, you can diagnose and treat potential crop issues before they become a widespread issue across the entire farm.

Spot spraying of crops used to be incredibly difficult. If you had an issue with weeds or a certain crop, the entire acreage would have to be sprayed.

This is a huge waste of time and resources, as someone will have to walk the entire acreage, plus there are the overall costs of pesticides and the associated environmental cost of chemical usage. With spot spraying afforded by drones, this same task can be accomplished in less time, with fewer monetary resources, and a reduced environmental cost.

Crop Mapping and Surveying

One of the biggest advantages of using drone technology is the ease and effectiveness of large-scale crop and acreage monitoring. In the past, satellite or plane imagery was used to help get a large scale view of the farm, while helping to spot potential issues.

However, these images were not only expensive but lacked the precision that drones can provide. Today, you can not only obtain real-time footage but also time-based animation which can illuminate crop progression in real-time.

With drone mapping and surveying, technology decisions can now be made based on real-time data, not outdated imagery, or best-practice guesswork. With near infrared (NIR) drone sensors you can actually determine plant health based upon light absorption, giving you a birds-eye view of the overall farm health.

With agriculture drones you'll be able to collect information like:

- The overall crop and plant health.
- Land distribution based on crop type.
- Current crop life cycle.
- Detailed GPS maps of current crop area.

The end result is simple, drones can help to maximize land and resource usage, and help farmers better determine crop planting locations.

Irrigation Monitoring and Management

Irrigation can be troublesome. With miles and miles of irrigation, issues are bound to arise. Drones that are equipped with thermal cameras can help to spot irrigation issues, or areas that are receiving too little or excessive moisture.

With this information, crops can be better laid out to maximize drainage, adhere to natural land runoff, and avoid water pooling, which can damage sensitive crops. Water and irrigation issues are not only costly but can ruin crop yields as well. With drone surveying, these issues can be spotted before they become troublesome.

Real-Time Livestock Monitoring

Some drones are equipped with thermal imaging cameras that enable a single pilot to manage and monitor livestock. This allows farmers to keep track of livestock a much greater frequency, and with less time and staff investment.

The drone operator can quickly check in on herd to see if there are any injured or missing livestock, as well as see livestock who are giving birth. Drones are used to keep an eye on the heard at all times, a once costly and time-intensive task.

Plus, thermal imaging will also help to keep an eye out for any livestock predators, which can be a huge advantage for some farm owners.

Applications of Agricultural Drone

Crop Monitoring

A professional UAV, such as the multispectral eBee SQ agriculture drone, provides a holistic view of a crop's growth, enabling ag professionals to quickly and precisely identify issues, and better target their field scouting. Multi-year drone data also allows the better planning and monitoring of improvements, such as ditches and evolving fertilizer applications.

Soil Assessment

Drone data can be employed to extract soil characteristics. These include temperature, moisture, slope, elevation and more. This ability enables more accurate soil sampling and the production of more suitable seeding prescriptions.

Plant Emergence and Population

Professionals like agronomists are in using the data from agricultural dro better understand which plants them well as population and spacing met information is then used to drive re decisions, thinning and pruning act to improve crop models.

Fertility

High-resolution drone data allows ag professionals to assess crop vigour at different stages of growth. This helps teams to apply just the right rates of fertiliser, reduce wastage and optimise crop health and production.

Crop Protection

The digital outputs generated by agricultural drones, when assessing stress and crop growth can help to guide the proper and efficient application of crop protection products. This allows for finely-tuned applications that meet the exact needs of each acre.

Insurance

On-demand, high-resolution drone perfect for capturing and accurately events that lead to economic loss, crop injury, destruction and reduce helping to create a more efficient a process.

Irrigation and Drainage

In addition to crop-specific activities such as health monitoring, agricultural drones equipped with RGB and/or thermal infrared cameras can also be used to plan and troubleshoot irrigation systems. This, in turn, helps professionals to manage their water flow and usage across an operation.

Harvest Planning

Drone data, collected at critical growth stages throughout the growing season, can help agronomists and ag engineers to improve their models, predictions and planning. The result is that teams can better anticipate both a harvest's quality and its final yield.

Pros and Cons of Drones in Agriculture

Pros

1. Analysis: Drones could be used for soil and field analysis. They can be used to produce accurate 3-D maps that can be used to conduct soil analysis on soil property, moisture content, and soil erosion. This is very important in planning seed planting patterns. Even after planting, such information is useful for both irrigation and the management of the nitrogen level in the soil.
2. Planting: Though not quite prevalent just yet, some manufacturers have come up with systems able to shoot pods containing seeds and plant nutrients into the already prepared soil. This profoundly reduces the planting costs.
3. Monitoring: One of the largest obstacles in farming is inefficient crop monitoring of vast fields. This challenge is made worse by the rise of unpredictable weather patterns which lead to increased risks and maintenance costs. Drones can be used to develop time series animations to show precise crop development which reveals production inefficiencies hence better crop management.

DJI Phantom 3 Professional surveying a field

4. Drones for Agriculture Spraying: Using ultrasonic echoing and lasers, drones can adjust altitude with a change in topography and geography. Their ability to scan and modulate its distance from the ground enables them to spray the correct amount of the desired liquid evenly in real time. This results in increased efficiency since the amount of water penetrating into groundwater is minimized. Spraying using drones has also proven to be faster than other traditional methods.

5. Irrigation: Drones equipped with thermal, hyper-spectral, or thermal sensors can identify the parts of the field have become dry. This way the identified areas can be attended to promptly making irrigation precise and timely.

6. Health Assessment: Some drones are capable of scanning crops using visible and near infrared light. On-board light processing devices are then able to identify the amounts of green and near-infrared light reflected by the plants. This data is then used to develop multi-spectral images which depict the plant health. These images can be used to track crop health and to monitor remedied administered if any sickness is discovered.

7. Ease of Deployment: Unlike traditional aircraft, the drones are easier and cheaper to deploy.

Cons

1. Flight Time and Flight Range: There are some problems with drones in agriculture. Most of the drones have a short flight time of between 20 minutes to an hour. This makes limits the acreage that it can cover for every charge. The flight range also limits the radius that can be cover during every flight time. Drones that can offer longer flight time and longer range are relatively costlier.

2. Initial Cost of Purchase: Drones with the features that qualify them for use in the agricultural are quite costly. This is mostly so for fixed wing drones which could cost up to $25000. For some drones, the heavy cost is inclusive of hardware, software, tools and imaging sensors.

   Buying drone that does not come equipped with the necessary equipment could be cheaper. However, the desired cameras and the processing software are quite costly making it equally capital intensive. Purchasing drones equipped for use in agriculture could prove costly in the short run but worthwhile in the long run.

3. Federal Laws: The use of drones for agriculture purposes is considered commercial. This means the farmer needs to undergo FAA operator training so as to acquire a remote pilot certificate or hire an operator with such qualifications. FAA also demands that drones be flown at an altitude of not more than 400 feet.

4. Interference within the Airspace: Agricultural drones share the same airspace with manually manned aircraft. Hence they are prone to interference. It's, therefore, advisable the farmer files his/her flight plan with the local airport or the FAA before the flight.

5. Connectivity: Most of the arable farmlands in the US have very little online coverage if any. This means that any farmer intending to use drones has to invest in connectivity or buy a drone capable of capturing and storing data locally in a format that can later be processed.

6. Weather Dependent: Drones are much more prone to climatic conditions, unlike traditional aircraft. If it is very windy outside or rainy outside you may not be able to fly them.

7. Knowledge and Skill: The images require analysis by skilled and knowledgeable personnel for them to translate to any useful information. This means an average farmer without these skills may need training or may be forced to hire skilled personnel conversant with the analysis software to help out with the image processing.

Drone technology keeps improving every day. With many manufacturers entering the industry, it's hoped that the cost of the drones and the accompanying equipment will reduce. Limitations like flight time and range are also expected to be solved by an improvement in technology. These improvements will ensure that farmers reap more from the use of drones.

References

- Mechanization-of-agriculture-meaning-benefits-and-progress, farming, india: economicsdiscussion.net, Retrieved 12 July, 2019
- Agricultural-technology: newworldencyclopedia.org, Retrieved 11 January, 2019
- Emerging-agriculture-technologies: ayokasystems.com, Retrieved 1 August, 2019
- Agriculture-present-applications-impact: emerj.com, Retrieved 5 March, 2019
- Machine-learning-in-agriculture-applications-and-techniques: medium.com, Retrieved 8 June, 2019
- How-Vision-Systems-are-Used-in-Agriculture: visiononline.org, Retrieved 3 May, 2019
- Agbot-agricultural-robot: techtarget.com, Retrieved 5 February, 2019
- Robotics-in-Agriculture-Types-and-Applications: robotics.org, Retrieved 30 April, 2019
- Agriculture-robots: postscapes.com, Retrieved 15 july, 2019
- Robots-that-are-invading-the-agriculture-industry: interestingengineering.com, Retrieved 30 August, 2019
- Agricultural-drones: uavcoach.com, Retrieved 16 January, 2019
- Agricultural-drones-industry, industry: sensefly.com, Retrieved 29 March, 2019
- The-pros-and-cons-of-drones-in-agriculture: droneguru.net, 20 April, 2019

2

Farm Machinery

Farm machinery is the machinery used in agricultural and farming activities. A few examples of such machinery are farm tractor, combine harvester, wind turbines, seed processing equipment, etc. The diverse applications of these machines have been thoroughly discussed in this chapter.

Precision Agriculture

Precision agriculture seeks to use new technologies to increase crop yields and profitability while lowering the levels of traditional inputs needed to grow crops (land, water, fertilizer, herbicides and insecticides). In other words, farmers utilizing precision agriculture are using less to grow more. GPS devices on tractors, for instance, allow farmers to plant crops in more efficient patterns and proceed from point A to point B with more precision, saving time and fuel. Fields can be leveled by lasers, which means water can be applied more efficiently and with less farm effluent running off into local streams and rivers. The result can be a boon for farmers and holds great potential for making agriculture more sustainable and increasing food availability.

Precision agriculture relies upon specialized equipment, software and IT services. The approach includes accessing real-time data about the conditions of the crops, soil and ambient air, along with other relevant information such as hyper-local weather predictions, labor costs and equipment availability. Predictive analytics software uses the data to provide farmers with guidance about crop rotation, optimal planting times, harvesting times and soil management.

Sensors in fields measure the moisture content and temperature of the soil and surrounding air. Satellites and robotic drones provide farmers with real-time images of individual plants. Information from those images can be processed and integrated with sensor and other data to yield guidance for immediate and future decisions, such as precisely what fields to water and when or where to plant a particular crop.

Agricultural control centers integrate sensor data and imaging input with other data, providing farmers with the ability to identify fields that require treatment and determine the optimum amount of water, fertilizers and pesticides to apply. This helps the farmer avoid wasting resources and prevent run-off, ensuring that the soil has just the right amount of additives for optimum health, while also reducing costs and controlling the farm's environmental impact.

In the past, precision agriculture was limited to larger operations which could support the IT infrastructure and other technology resources required to fully implement and benefit from the benefits of precision agriculture. Today, however, mobile apps, smart sensors, drones and cloud computing makes precision agriculture possible for farming cooperatives and even small family farms.

Controlled Traffic Farming

Controlled traffic farming (CTF) is all about managing soil compaction – confining it to narrow strips across the land and maximizing the remaining undamaged soil area for cropping. In practice it means matching machinery tracks so they take up the least possible area. Although this is made simpler by satellite guidance, it can be achieved with conventional marking systems. Farm conversion to CTF in the first instance means adopting a CTF "mindset" – the belief that separating wheels and crops is a key method of reducing costs and increasing returns. From here on it is simply a matter of good planning and timely investment that ensures a minimum 15% return on capital, an increase in crop returns and a substantial reduction in costs.

Controlled traffic farming cuts fuel, labour and machinery costs by reducing soil damage. This makes farming simpler, more reliable and less time consuming. It also delivers environmental benefits, such as reduced water run-off and soil erosion, improved fertilizer use efficiency, less risk of nitrous oxide and methane emissions and improved carbon sequestration. Overall, with reduced fuel, energy and machinery inputs and fewer greenhouse gas emissions, the carbon footprint of CTF is likely to be the lowest of all farming systems.

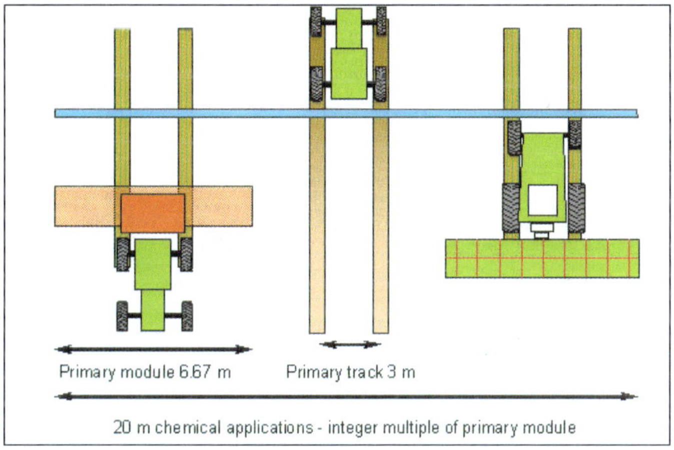

CTF system with base module of 9 m and a 3 times multiple to give 27 m for chemical applications. The wheel track width is 3 m

CTF is a simple way of reducing input costs (time, fuel and machinery) – and at the same time increasing crop yields – both of which are done sustainable and both of which increase farm profit. CTF is a whole farm approach to the separation of crops and wheels; it is a system that avoids the extensive soil damage and costs imposed by normal methods. Controlled traffic is not rocket science it simply involves confining all field vehicles to the least possible area of permanent traffic lanes. Appropriate agronomy and management is used to maximize the potential of both the cropped and wheeled areas for their specific purposes. In practice it means the repeated use of the same wheel tracks for every operation, and although it is ideal for all machines to have the same wheel track (the distance between the left and right wheel centres) and for all implements to have a particular span (base module) or whole number multiple of it, this is not essential. The percentage area wheeled can be reduced to 30-40 % even with two different track and implement widths. The illustration shows an optimized CTF set up with planter, harvester and chemical applicator.

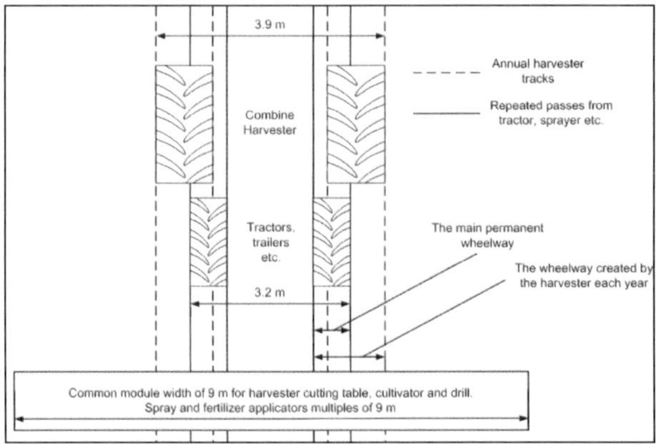

Common module for different tyres of tractor, sprayer, trailer and harvester

CTF is appropriate for anyone growing crops, whether these are grass, roots, energy, legumes or cereals, on any scale and whether with manual, semi- or highly-mechanised systems. It is a system that cuts costs at their source; it creates opportunities and avoids compromises associated with wheels, rutting and poor cloddy seedbeds. It should be the goal of all producers. CTF is possible to be used in conventional soil tillage system based on ploughing, but ploughing removes some of the benefits and the hope is that it will not be necessary. Obviously ploughing will loosen all of the wheel ways and in many cases displace the compacted soil laterally. To minimise the damage caused by ploughing, the tractor hauling the plough should work on the land, not in the furrow, and the direction of ploughing should be at right angles to the wheel ways.

There are many benefits associated with CTF and they all help to deliver the two most important factors in farming operations increased profit and improved sustainability. These are delivered by improving soil health, which in turn lowers costs and increases crop returns but it also results in improved environmental conditions. Lower costs and increased returns are brought about by:

- Lower energy for cultivation.
- Lower energy for driving over the soil.
- Lower machinery investment.
- Better seedbeds.
- Improved soil structure.
- Increased potential and accuracy for global positioning systems.
- Improved fertilizer use efficiency.
- Potential to retain more organic matter and soil living organisms.
- Improved water storage.

Converting to CTF does need some thought and once achieved, more discipline. No longer must growers think of their fields as open areas to sneak across when in a hurry; the soil becomes sacrosanct and is treated with extreme care. There is a perception that converting to CTF is costly but the reality is large savings in time, energy, labour, fuel and machinery investment. The more thought and the better the planning, the greater will the savings be. The first stage in conversion is a machinery inventory – what have you got and how does it measure up and how much of it will you use in the future when your soil may need little or no cultivation? These deliberations together with increasingly sophisticated tools to help growers make the transition, means that a CTF system customised to each farm will present itself.

Because machines have not so far been designed with CTF in mind, there will be constraints, but a 25 % tracked area is vastly different from annual 75% random tracking. This is because damage takes many years to repair naturally, particularly deeper in the profile. Another way of reducing the tracked areas is to choose narrower but perhaps larger diameter tyres, making sure they can take the loads imposed on them. These narrower tyres are potentially less costly and it therefore pays to buy the best quality, but only experience will tell you how narrow in terms of maintaining the traffic lanes.

Autonomous Farm Equipment

Autonomous equipment technology will tackle the growing concern of labor shortages while boosting productivity and ecomonical savings, which will bring higher yields. The technology will allow for true 24/7 operations and a farm manager will be able to manage equipment operation from a home office or on a tablet while on the go.

CASE IH Case autonomous tractor

Each autonomous farm vehicle is equipped with a series of hardware and software components that permit a user to switch between manual and robotic control. The components work with the existing mechanical, or hydraulic control system and link the vehicles to a central control station, which will permit a single operator to manage multiple vehicles at the same time throughout the farm's operation.

A study reported that 71% of labor-intensive specialty crop growers experience labor shortages. Farm automation technology will help surmount paralyzing labor shortages while increasing productivity and safety.

But, are fully autonomous farm equipment viable? Fully autonomous tractors require different and often overlapping sensors to navigate properly to avoid collisions. The technology is ready at the prototype level, and the components required are ready, such as lidar, sonar, and radar. Nevertheless, the autonomous equipment will require several more years of actual field trials. Components such as lidar are fairly expensive.

It will take time to address other issues such as liability and insurance. Regulatory issues could prevent agriculture from moving forward with this technology. Farmers need to know their liability. agBOT teams are working to make self-regulation recommendations on autonomous vehicles for farm safety.

Although autonomous farm vehicles are effective, farmers still prefer to remain in charge. Therefore, these vehicles must be 100% reliable before a farmer will place trust in them. So, what is being done? Machine vision is making robots more adaptable than ever. Machine vision is a crucial element to robotic systems that perform effectively in agricultural environments, because careful and controlled movements are routine, but not identical.

The John Deere AutoTrac system is an innovative step forward for precision planting. Using enhanced GPS, it is able to plant lines of crops with a minimum distance of about one inch between each line. The overlap between tilling, planting, and fertilizing can be reduced, thereby reducing chemical treatment and operator fatigue.

Robots can harvest crops and accomplish many other manual tasks. They can also support farming in new, experimental directions. Using machine vision cameras, robots can accurately measure plant characteristics. This allows the robots to develop databases central to making genetic improvements in crops.

Agricultural drones are the eye in the sky that permits farmers to evaluate the status of crops. Problems such as pests, disease, and the effects of weather can be analyzed swiftly through a combination of low and high-flying images supported by precision machine vision sensors.

Machine vision is becoming cutting-edge enough to introduce totally autonomous robots into many facets of agriculture. As AI, machine vision, and motion control progress, fully autonomous farms may be achieved. This could be an important step to provide for an ever-growing population.

Drones or unmanned aerial vehicles are projected to bring an increase in the autonomous farm equipment industry. Autonomous Drones will be used for the purpose of crop monitoring, spraying, planting and field analysis. Drones can reach any location in the field quickly and can efficiently recognize problems, using scanners, thermal sensors, cameras, ultra-sonic echo machines, and 3D maps.

Drones can be helpful at the beginning of the crop cycle. They produce detailed 3-D images for early soil analysis, and are valuable in planning seed planting patterns. After planting, the analysis from the drone provides data for irrigation and nitrogen-level management.There are drone planting systems that accomplish a realization rate of 75 percent and reduce planting costs by 85 percent. The drone shoots seed pods with plant nutrients into the soil, delivering all the nutrients necessary for the plant to support life.

Drones can scan the ground and spray the precise amount of chemicals, spraying in real time for uniform coverage, resulting in improved efficiency with a lessening of the amount of chemicals entering the groundwater.

Crop monitoring challenges are worsened by the increase of unpredictable weather conditions, which drive up chance with higher field maintenance costs. A drone can show the detailed development of a crop and expose production inefficiencies, allowing for better crop management.

A farming robot revolution is impending, with fruit-picking machines ready to roll into the fields and replace human workers. Whether it's berries, lettuce or grapes, the farmers are jostling for available labor.

Pitfalls of the Driverless Tractor

Innovators are often short sighted. They often want new technology to do exactly what the old technology did, only better and cheaper. They are often slow to realize the potential of the new technology to do things that the old could not. For example, early tractor manufacturers wanted a machine that was stronger than a team of horses and cheaper to keep, but which would essentially perform the same job as flesh and blood horsepower. It took them some time to realize that they were not bound by the limitations of animal traction that could provide only draft pulling power and turning power through ground drive wheels. Tractors not only had more power, but more easily controllable mechanical power (e.g., power take off (PTO), hydraulics). Many of the cropping systems developments of the 20th century would have been much more difficult if they had to depend on the power from ground drive wheels. For example, no-till planting is possible with horses (some of the Amish do it), but it is doubtful if it would have ever been invented without the power of tractors and the ease of hydraulics.

One of the next innovations in the precision farming line-up may be falling prey to this pitfall. Reports of autonomous tractors and other equipment are starting to appear in the farm and business press. Usually, the description is of a conventional machine where the driver has been replaced by some combination of computers, global positioning system (GPS) and electronic sensors. The benefits of this approach usually focus on higher productivity because:

- The computer does not get sick or tired and does not need time off,
- It can operate with closer tolerances (so every round is at full field capacity), fewer errors and at higher speeds, and
- Because machines can be made lighter and cheaper if the drivers seat, controls and cab can be eliminated.

Legal Liability

One of the key disadvantages of driverless machines for agriculture is liability. Unlike factory robots, agricultural machines must work in public. One news report of a malfunctioning machine that crashes into a neighbor's yard or of a machine that fails to recognize a dog or child and runs it over would create a firestorm of negative publicity. This type of accident is not new. Unfortunately, every year pets and children are hurt by tractors and other equipment. What is different for autonomous equipment is the perception that the accident was in part due the lack of a human to intervene.

If autonomous equipment becomes commercially available, insurance companies would probably insist that a driver be present, if only to shut off the switch in case of a malfunction. This reemphasized that a driver would greatly reduce the benefits of this technology. Even if the driver is only present for safety reasons, he or she can still get sick, tired or need time off. If the driver is reinstated, a seat, cab and controls must be provided.

The liability issue is really an aspect of community acceptance of this technology. Community acceptance is an issue with all new technology and tends to be more important where the new technology forces major changes in long established ways of life and/or seems to present some physical danger. Not everyone greeted the arrival of automobiles and tractors with open arms. Motorized vehicles scared horses. Some people thought they were a safety hazard. Horse breeders quickly recognized that motorized transport would put them out of business. Today, new hog facilities are very efficient in turning corn into pork. Financing is available. The greatest challenge that investors who wish to build new hog facilities often find themselves facing is convincing the surrounding community to let them build.

Would people eventually become accustomed to autonomous farm equipment? Over time, they would probably become more accustomed, but perhaps only after insisting on a variety of safety measures. In addition to a rider to hit the off switch, there might be legal requirements that fields farmed with remote control equipment be fenced to prevent children from straying into the path of the equipment. This would make the driverless equipment in the field more like the robot in the factory. For high value crops, such as the biotech pharmacy crops, extra safety measures might be affordable, but for bulk commodities, they may drive the cost out of reach.

The other problems associated with autonomous farm equipment can probably be overcome with technology. Better sensors and controls would allow the equipment to deal with plugging and malfunctions on its own. In addition to operating equipment, drivers are also collecting information (e.g., weed, disease and insect problems, soil issues, stand establishment). If they are no longer going across the field regularly, other ways need to be found to collect non-standard data. Better sensors would help. Improved scouting programs would be essential. Nevertheless, we will never have a sensor for every possible problem; a periodic human presence in the field is likely to be necessary for the near future.

Alternative Paradigm

To a large extent, the liability and safety issue is one of perception. It is generated by a concern for large equipment working in public with no visible human guidance. Size is an issue here. A lovable R2-D2 type robot working in the field would generate far less concern than a remote controlled 4WD tractor with a 40-foot implement traveling at 10 mph. The irony of this observation is that once the driver is removed, bigger farm equipment is no longer obviously better. One of the main advantages of motorized mechanization was that it allowed one person to accomplish more. The more power the driver could control and the wider the equipment, the more one person could do in a given time and the lower the labor cost. With computer control, one person could supervise a swarm of smaller machines just as easily as one person could supervise one large one. The ideal autonomous equipment may be more like a rototiller than a 4WD tractor.

Imagine an agricultural system in which most physical work is done by small robot like machines, "droids" if we want to use the Star Wars terminology. Each field could have one or more depending on its size. The machines could stay in the field for the whole season stored in a shed in the corner of the field and set in motion by commands from the farm office, or perhaps from a computer in the pickup.

For very small fields (e.g. in suburbanized areas), the machines would be small enough to load in the back of a pickup. A common farm management chore would be to bring the machines out to new fields and set them to work.

The optimal area that each machine would cover at one pass is a technical question that would probably depend on the crop and the equipment design, but for row crops it is possible that the machine would work on one or two rows at a time. The machine does not tire or need breaks, so it could work day and night giving close attention to each row. This could be the organic farmer's dream, essentially allowing gardening on a mass scale. Weeds could be pulled or hoed. Insects picked off one-by-one and crushed.

This type of technology would have enormous implications for farm operations. Imagine planting time on a large mid-western farm. One farm operation may have hundreds (or thousands) of machines planting simultaneously. They would take seed, fertilizer and pesticides (if they are still used) from holding bins in their fields. Human intervention is needed in several areas: keeping the machines running, supplying them with inputs and especially deciding what the machines should be doing.

The technology also has implications for the value of farmland. The productivity of the soil would continue to be very important, but the size and shape of the fields would become less important. The machines could easily work around trees, rocks, ponds and other obstacles. Small suburban fields could be worked almost as efficiently as large tracts of land.

Farm Tractor

Tractor is a self-propelled power unit having wheels or tracks for operating agricultural implements and machines including trailers. Tractor engine is used as a prime mover for active tools and stationary farm machinery through power take-off shaft (PTO) or belt pulley.

Classification of Tractors

Tractors can be classified into three classes on the basis of structural-design:

1. Wheel tractor: Tractors, having three of four pneumatic wheels are called wheel tractors. Four wheel tractors are most popular everywhere.
2. Crawler tractor: This is also called track type tractor or chain type tractor. In such tractors, there is endless chain or track in place of pneumatic wheels.
3. Walking tractor (Power tiller): Power tiller is a walking type tractor. This tractor is usually fitted with two wheels only. The direction of travel and its controls for field operation is performed by the operator, walking behind the tractor.

On the basis of purpose, wheeled tractor is classified into three groups:

1. General purpose tractor: It is used for major farm operations; such as ploughing, harrowing, sowing, harvesting and transporting work. Such tractors have:
    - Low ground clearance
    - Increased engine power

- Good adhesion
- Wide tyres

2. Row crop tractor: It is used for crop cultivation. Such tractor is provided with replaceable driving wheels of different tread widths. It has high ground clearance to save damage of crops. Wide wheel track can be adjusted to suit inter row distance.

3. Special purpose tractor: It is used for definite jobs like cotton fields, marshy land, hillsides, garden etc. Special designs are there for special purpose tractor.

Tractor Components

A tractor is made of following main components:

1. I. C. engine,
2. Clutch,
3. Transmission gears,
4. Differential unit,
5. Final drive,
6. Rear wheels,
7. Front wheels,
8. Steering mechanism,
9. Hydraulic control and hitch system,
10. Brakes,
11. Power take-off unit,
12. Tractor pulley and
13. Control panel.

Every tractor is fitted with an I. C. engine, the engine may be carburettor type or diesel type but nowadays almost all the tractors are diesel tractors.

Selection of Tractor

Selection of tractor depends upon following factors:

1. Land holding: Under a single cropping pattern, it is normally recommended to consider 1 hp for every 1 hectares of land, In other words, one tractor of 20-25 hp is suitable for 20 hectares farm.

2. Cropping pattern: Generally less than 1.0 hectare/hp have been recommended where adequate irrigation facilities are available and more than one crop is taken. So a 30-35 hp tractor is suitable for 25 hectares farm.

3. Soil condition: A tractor with less wheel base, higher ground clearance and low overall weight may work successfully in lighter soil but it will not be able to give sufficient depth in black cotton soil.
4. Climatic condition: For very hot zone and desert area, air cooled engines are preferred over water-cooled engines. Similarly for higher altitude, air cooled engines are preferred because water is liable to be frozen at higher altitude.
5. Repairing facilities: It should be ensured that the tractor to be purchased has a dealer at nearby place with all the technical skills for repair and maintenance of machine.
6. Running cost: Tractors with less specific fuel consumption should be preferred over others so that running cost may be less.
7. Initial cost and resale value: While keeping the resale value in mind, the initial cost should not be very high; otherwise higher amount of interest will have to be paid.
8. Test report: Test report of tractors released from farm machinery testing stations should be consulted for guidance.

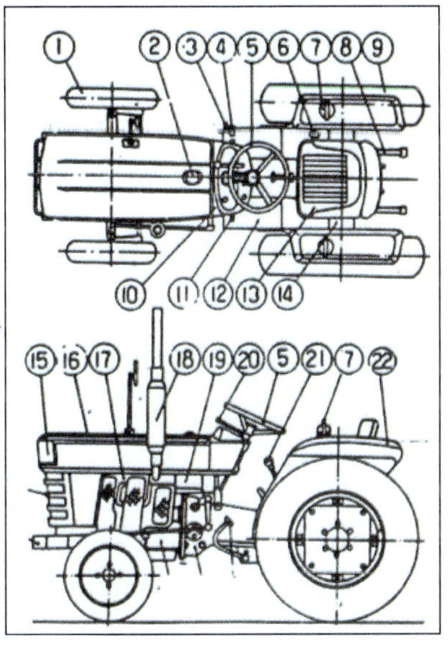

Componenets of Tractor

In above figure,

1. Front wheel,
2. Fuel tank cap,
3. Accelerator pedal,
4. Brake pedal,
5. Steering wheel,
6. Hydraulic control lever,
7. Turn signal lamp,

8. Lift arm,
9. Rear tire,
10. Instrument panel,
11. Clutch pedal,
12. Step,
13. Seat,
14. Rear axle housing,
15. Side clearance,
16. Engine hood,
17. Fan cover,
18. Muffler,
19. Fuel tank,
20. Throttle lever,
21. Main speed change lever,
22. Fender.

Control Board or Dash Board of a Tractor

The control board of a tractor generally consists of:

1. Main switch,
2. Throttle lever,
3. Decompression lever,
4. Hour meter,
5. Light switch,
6. Horn button,
7. Battery charging indicator,
8. Oil pressure indicator,
9. Water temperature gauge.

Tractor Tyres and Front Axle

- Tyres: The tyres are available in many sizes with the ply ratings as 4, 6 or 8. The ply rating of tyres indicates the comparative strength of tyres. The higher the rating, the stronger are the tyres. The tyres size 12—38 means, that the sectional diameter of tyres is 12" and it is mounted on a rim of 38" diameter. The inflation pressure in the rear wheels of the tractor varies between 0.8 to 1.5 kg/cm². The inflation pressure of the front wheel varies from 1.5to2.5kg/cm². Useful life of the pneumatic tyres under normal operating condition may be about 6000 working hours for drawbar work.

- Front Axle: Front axle is the unit on which front wheel is mounted. This wheel is an idler wheel by which tractor is steered in various directions. The axle is a rigid tubular or I-section steel construction pivoted at the centre. There are various adjustments of front wheel.

Hitching System of Tractor Drawn Implements

Tractor drawn implements possess higher working capacity and are operated at higher speeds. These implements need more technical knowledge for operations and maintenance work. Tractor drawn implements may be: a) Trailed type b) Semi-mounted type and c) Mounted type.

a. Trailed type implement: It is one that is pulled and guided from single hitch point but its weight is not supported by the tractor.

b. Semi-mounted type implement: This type of implement is one which is attached to the tractor along a hinge axis and not at a single hitch point. It is controlled directly by tractor steering unit but its weight is partly supported by the tractor

c. Mounted type implement: A mounted implement is one which is attached to the tractor, such that it can be controlled directly by the tractor steering unit. The implement is carried fully by the tractor when out of work.

Some Important Terms Connected with Tractors

- Wheelbase: Wheel base is the horizontal distance between the front and rear wheels of a tractor, measured at the ground contact.
- Ground clearance: It is the height of the lowest point of die tractor from the ground surface, the tractor being loaded to its maximum permissible weight.
- Track: Track is the distance between the two wheels of the tractor on the same axle, measured at the point of ground contact.
- Turning space: It is the diameter of the smallest circle, described by the outermost point of the tractor, while moving at a speed, not exceeding 2 km/hr with the steering wheels in full lock.
- Cage wheel: It is a wheel or an attachment to a wheel with spaced cross bars for improving the traction of the tractor in a wet field. It is generally used in paddy fields.

Power Tiller

It is a prime mover in which the direction of travel and its control for field operation is performed by the operator walking behind it. It is also known as hand tractor or walking type tractor. The concept of power tiller came in the world in the year 1920. Japan is the first country to use power tiller on large scale. In Japan, the first successful model of power tiller was designed in the year

1947. Production of power tiller rapidly increased during the year 1950 to 1965. Power tiller was first introduced in India in the year 1963.

Power Tiller

Power tiller is a walking type tractor. The operator walks behind the power tiller, holding the two handles of power tiller in his own hands. Power tiller may be called a single axle walking type tractor, though a riding seat is provided in certain designs. Average size of holding in India is about 2.5 hectares. There are 89% of total land holdings of less than six hectares. Under such conditions, power tiller may be useful as a power unit.

Components of Power Tiller

A power tiller consists of the following main parts:

1. Engine
2. Transmission gears
3. Clutch
4. Brakes
5. Rotary unit

All the power tillers are fitted with an I. C. engine. At present, most of the power tillers are fitted with diesel engine. The makes like Kubota, Mitsubishi, and Sarachi have used diesel engine in India.

Operation

The main clutch is a lever on the handle. The lever can be shifted to on or off position while operating in the field. When the lever is shifted to on position, the power from the engine is transmitted through the main clutch to the various parts of the power tiller. When the lever is shifted to off position the power from the engine is cut-off from the rest of the transmission.

Power Transmission in Power Tiller

For operation of power tiller, the power is obtained from the IC Engine, fitted on the power tiller. The engine power goes to the main clutch with the help of belt or chain. From main clutch, the power is divided in two routes, one goes to transmission gears, steering clutch and then to the wheel. The other component goes to the tilling clutch and then to the tilling attachment.

The flow diagram for transmission of power is given below:

Engine → Main clutch → Transmission gear → Steering clutch → Wheels

Engine → Main clutch → Tilling clutch → Tilling attachment

V-belt is usually used to transmit power from the engine to the main clutch, because V-belt has very high efficiency and it works as a shock absorber also.

Main clutch - Power goes from the engine to the main clutch. Clutch may be:

1. Friction clutch or
2. V-belt tension clutch.

Friction clutch is generally used for bigger power tiller. Usually it is a dry type multiple disc clutch. V-belt tension clutch is used for small power tillers. The main functions of clutch in a power tiller are:

1. To transmit engine power to transmission gears and
2. To make power transmission gradual and smooth.

1. Transmission gears: Transmission box consists of gears, shafts and bearings. The speed change device may be; (a) gear type or (b) belt type.

2. Brakes: All power tillers have some braking arrangement for stopping the movement. Most of the power tillers use inner side expansion type brake.

3. Wheels: Usually 2 to 4 ply pneumatic tyres are used in power tillers. The pressure of the tyre ranges from 1.1 to 1.4 kg/cm^2.

4. Rotary unit: Power tiller has a rotary unit for field operation. Rotary unit is of two types: (a) Centre drive type and (b) Side drive type.

Centre drive type has got transmission at the centre and the side drive type has transmission at one side. Centre drive type has the following characteristics: (a) Tilling width can be widened (b) Rotary unit is light in weight (c) Fixing of attachment is easy (d) The tine shaft can be detached easily (e) Mounting and dismounting of rotary unit is very easy (f) It may leave some portion of the field untilled (g) It has one point support on the ground.

In side drive type - (i) Deeper tilling is possible (ii) The arrangement is useful for hard soil (iii) It has two points support on the ground.

1. Rotary tines: Rotary tines are used in rotary unit for soil cutting and pulverisation purpose. Rotary tines are of three types: (i) Straight tines (ii) Curved tines and (iii) Sliding tines.

   In case of straight tines: (a) Power consumption is less (b) Fine pulverisation of soil is possible (c) Poor soil turning (d) Grass entangles in the tines very easily (e) It is suitable for hard soil. In case of curved tines: (a) Good soil turning is possible (b) It is suitable for avoiding grasses (c) Pulverisation of soil is coarse and (d) Power consumption is high.

Sliding tines have the characteristics of sliding on their positions according to the requirement.

2. Steering cutch lever: Steering clutch is provided on the grip of the right and left handles. When (he left side is gripped, power is cut-off on left side of the wheel and the power tiller turns to the left. Similarly when the right side is gripped, the power tiller turns to the right.

Autonomous Tractors

Agribusinesses today face a variety of challenges, not the least which are Mother Nature's excesses, depressed commodity prices, expensive and complex equipment, high energy costs and the lack of qualified and eager help. Autonomous, or operator-less technology, increasingly found in material handling systems used in factories and slowly finding acceptance in automotive vehicle lane keeping assist systems—not to mention beta self-driving models from manufacturers such as Tesla and Google – offer varying degrees of reduced operator input.

Traditional GPS guidance systems work well within their reception constraints, but when interferences such as from power transmission lines, sunspots and other dead zones are encountered, are just not sufficient for autonomous operation.

Autonomous Tractor Corporation (ATC), headquartered in Fargo, ND, has introduced AutoDrive, a patented navigation system that combines GPS technology with two individual on-ground validation systems. At the heart of AutoDrive is ATC's Laser-Radio Navigation System (LRNS), a Local Area Network (LAN) system based on Radio Frequency (RF) technology that utilizes artificial intelligence (IA) to "train" the tractor to do repetitive tasks without programming.

Much as farmers introduce sons and daughters to tractor operation by having them first sit alongside and observe for a time, then slowly moving to where the youth, under close supervision, performs a single task, AutoDrive is trained to perform the task just as the farmer does. As in the case of the youth, when AutoDrive becomes proficient at that task and suitably trusted, slowly more tasks can be introduced. The parent farmer has no more trust for an autonomous tractor then for a son or daughter until a certain level of trust is developed over time. For safety, AutoDrive comes with a sonar-based detection system that will immediately bring the tractor to a halt when an object is detected 30' away. When this occurs, a text is sent to the owner, who through two pan-tilt cameras can remotely access what has happened.

This type of drivetrain is also complex with many components and moving parts, mandating a high degree of required maintenance that only increases as it ages. studies have shown that 25% of tractor maintenance gets done in years one through seven, but that 75% is required in years seven through to the expected life of fifteen years. These traditional drivetrains are also not the perfect candidates for autonomous operation. Electronically controlled systems, as opposed to mechanical ones, are safer and easier to control and use.

The drivetrain that ATC has developed is a perfect complement to AutoDrive and utilizes electric motors to provide power. ATC calls its system eDrive. Though electric motors provide instant torque, have long life and are ideal for electronic control, they do have low speed and heat-related drawbacks. To overcome those obstacles, the electric motors ATC has developed are fully liquid-cooled and internally lubricated (no external grease fittings) and have proprietary electronic controls that allow constant low-speed operation without sacrificing motor life. Beginning with heavy wall pipe machined to tight tolerances, a stator/shaft assembly is inserted along with the necessary electronics, and then the shell is filled with oil (the motor also utilizes an external reservoir) and then capped off. The completed motor is sturdy enough to serve as axle and drivetrain and in the case of a large tractor, such as the John Deere 8760 articulated machine utilized as the first test rig, is located at each drive wheel (100HP x 4). This first machine became known as JD400e providing 400 gross horsepower, 352 drawbar horsepower, a top transport speed of 20 mph and 32,145 lb-ft of torque at 4mph – all at half the price of new.

For smaller tractors (ATC is currently retrofitting a John Deere 6415 four-wheel drive tractor) where there is not enough space to have individual motors at each wheel, one or two motors driven off the engine are used. This allows flexibility, such as using a front drive motor providing a higher speed (transport) along with a rear drive motor that provides more torque (implement demand). Currently ATC is retrofitting 100, 200, and 400-horsepower tractors with eDrive technology, and though the first machines have been John Deere, the technology is compatible with any manufacturer's machines.

Taking full advantage of ATC's electric motor and control technology, the Spirit AT400 was developed. Spirit is a fully autonomous tractor based on a skid-steer platform that does not have a cab – much less an operator cockpit. Spirit is a natural progression of applying autonomous technology: the future holds even more. Imagine not having to have a tractor power the implement but rather having the drive motors onboard the implement and a diesel gen-set as a transferrable power source that could move from implement to implement.

Grain Drills and Planters

Grain drills and seed planters are key components in developing successful conservation plantings. They are complex machines that deliver seed at a metered rate, place it at a consistent depth in the soil, and produce light compaction to provide good seed to soil contact. Planters and drills come in many different forms with varying strength and weaknesses depending on the seed being used and condition of the planting site. Some require prepared seed beds, others require little to no seed bed preparation, and others are capable of preparing the seed bed and planting in a single pass. Understanding the basic operation of these machines and their strengths and weaknesses is critical in determining which planter is best suited to meet the desired goals of the planting, or getting the best performance from the available planter.

Basic Design and Operation of a Grain Drill and Planters

Grain drills and planters, regardless of type, operate in the same basic fashion. Seed is held in a box while a mechanism driven by the ground wheels or disks drops seed at a metered rate. Seed falls to the soil surface where some form of compaction seals it in the ground. Seed boxes come in three configurations, standard, legume, and native grass (fluffy) seed. Each box is designed to handle specific seed and can usually be calibrated independently of one another. This attribute is very helpful when planting mixes of seed to achieve the desired rate of each species.

Side view of a no-till drill with a 3 seed box configuration for legumes, native grass, and standard seed. The size difference of the legume seed box, first box on left.

Standard seed boxes are designed to handle large smooth seed similar to corn, soybeans, peas, wheat, oats, etc. They may or may not have agitation to keep seed moving. They generally rely on the smooth nature of the seed and gravity to feed the planting mechanism. These seed boxes will not handle light, fluffy seed. Seed will bridge on itself and not feed properly; clogging the planter. It is also difficult to calibrate with tiny seed such as clovers. Small seeded species require a smaller delivery mechanism to achieve the correct planting rates. Seed is typically delivered from the seed box through drop tubes to a set of disc shaped "V" openers where it falls into a slice in the soil left by the openers before being covered and sealed with some form of compaction device.

The inside of a standard seed box showing the gravity fed mechanism that delivers seed to drop tubes. Seed is metered out via rollers with gear like teeth in the openings that control the rate seed flows to the drop tubes.

The legume box is designed to handle extremely small seed, and is much smaller in size than the standard seed box. Operation is generally the same and relies on gravity to feed the mechanism delivering seed to the drop tubes. The drop tubes deliver seed to the ground surface just in front of a compaction device such as press wheels or a cultipacker. Seed is planted shallower than seed from the standard seed box. Large seed can jam the smaller delivery mechanism of the legume box and should not be used.

The smaller seed delivery mechanism of a legume seed box on a no-till drill. Large seed will jam the small gears that deliver seed to the drop tubes

The native, or fluffy seed box is similar to the standard seed box, but has some form of aggressive agitation within the box to keep seed moving so it doesn't bridge on itself. Many will have a "pick" wheel mechanism that reaches up into the seed box and pulls seed down into the machine to deliver it to the drop tubes. The drop tubes are much larger in diameter to help prevent seed from bridging. Removing the fluffy awns and appendages of native grass seed, a process called debearding, will greatly reduce seed bridging in the seed box and drop tubes. The use of a dry seed lubricant such as powdered graphite is also recommended to aid seed flow.

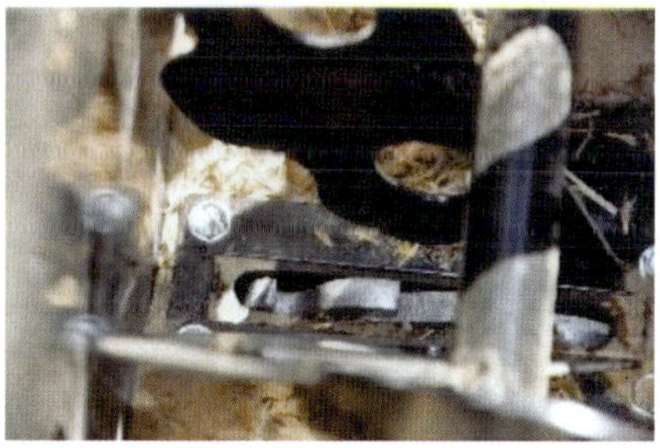

Close up view of the agitation auger and the pick wheel inside a native seed box.

The gear like teeth on the pick wheel reaches into the seed box, grabs seed, and pull it down into the drop tubes at a steady rate rather than relying on gravity to feed the mechanism. The auger blades move seed back and forth to keep it constantly moving and prevents seed from bridging on itself.

Standard Grain Drill

Standard grain drills are typically the most common planters available. These planters place seed in narrow rows approximately 6 to 8 inches apart. They are most commonly used for seeding pastures and planting cereal grain crops such as wheat, oats, and rye. These machines most often use a pair of discs called "V" openers to slice the soil, creating a shallow furrow for the seed to be placed. Press wheels follow this action and use light compaction to seal the opening created by the "V" openers. Press wheels are spring loaded to allow adjustment of down pressure based on the soil conditions. They require a prepared seed bed for proper operation. This would include some form of tillage to break the ground, smoothing to break up large clods and clumps of soil, and dragging or cultipacking to create a firm seed bed. Because they require prepared seed beds, they may not meet the needs and goals of soil health and conservation plantings where minimal soil disturbance is desired.

A grain drill with a standard seed box, note the V openers and press wheels with no fluted coulter.

No-till Drills

No-till drills operate in a similar fashion to standard grain drills, and also place seed in narrow rows. Seed from the standard and native seed boxes are delivered to an opening created by "V" openers and sealed with a set of adjustable press wheels. The major difference between No-till drills and standard grain drills is they do not require a prepared seed bed. Planting sites should be mowed and or treated with a broad spectrum herbicide before planting to reduce competition. They also differ by having a fluted, straight disc or coulter that slices through the unprepared seed bed creating a cut for the "V" openers to follow and open for seed placement. These planters are typically much heavier in construction than standard grain drills. The extra weight helps them break through existing plant residue, sod, or hard soil crust. They may require a larger tractor to pull and lift. Because they create virtually no soil disturbance, they are favored for soil health and conservation plantings where minimal change to the soil structure is desired. This minimal disturbance to the soil also reduces weed competition from early successional annual species that are set free by soil disturbances during typical seed bed preparation such as disking. Planting depth can be difficult to control with some no-till drills, especially with small seeded species because of the planter's increased weight and the opening created by the fluted cutting disc. This problem is compounded if the planter's drive mechanism is driven by the V-openers or fluted coulter disk. In order for this mechanism to operate consistently, it has to have firm contact with the soil, sometimes cutting too deep for small seeded species. No-till drills that operate their drive mechanism via rubber ground wheels with adjustable height reduce, but don't eliminate this problem. Depth bands on the coulter and or "V" openers also help reduce planting depth problems.

Front view of a no-till drill showing the rubber drive wheel, center. This planter has an optional research plot option attached to the top; not typical of standard drills used in conservation plantings.

The front fluted coulter of a no-till drill slices an opening through hard sods for the "V" openers, seen middle, to open and deposit seed. The spring loaded press wheels then seal the opening.

Specialty Drills and Planters

Drop Seeders

Drop seeders are similar to standard grain drills, but do not plant in neatly spaced rows or use "V" openers. These planters drop seed along the full length of the seeding box where it then falls randomly between a set of cultipackers and is pressed into the ground. Rolling cultipackers are used to drive the planter mechanism. They work well on prepared seed beds, and are capable of smoothing, planting, and cultipacking in a single pass; reducing some of the site preparation steps associated with a standard grain drill. They are also used to over seed small seeded species such as clover in existing pasture grasses. They handle small seed such as legumes and hard seeded grasses like Bermudagrass very well, and eliminate the issue of planting small seeds too deeply. They will not plant large seeded species deep enough to insure a reliable stand. They are rarely equipped with native seed boxes, unless specifically ordered.

A typical drop seeder with two sets of cultipackers. Seed falls from the seed box between these rolling cultipackers during planting. These are typically used with small seed.

Food Plot and all in one Planters

Ever changing improvements in land management for wildlife have created new hybrid types of planters that combine attributes of the seed bed prepping implements, grain drills, and drop seeders in one machine. They typically have a tillage device in front of a seed box that drops seed on the tilled seed bed in front of a set of cultipackers. They are capable of preparing the seed bed, planting, and cultipacking in a single pass; reducing time spent in the field and fuel used. They often come with a standard and legume seed box, and some brands have native seed boxes available upon request. Some models have short drop tubes which can be moved to multiple locations to help with planting depth. Large seed can be set to fall directly behind the tillage device, while small seed can be delivered between the cultipackers for a shallower planting depth. These planters are typically small 4 to 7 feet in width for use in wooded areas and small food plots. Although they do create soil disturbance to develop a seed bed; the disturbance is shallow, typically less than 3 inches. Models with disks to create the seed bed often have a mechanism to alter the angle of attack of the blades so the user can create as much or little disturbance as desired. The Brillion Till 'n Seed uses a set of counter rotating sprockets to create a very shallow seed bed, less than an inch in depth, which may meet the minimal soil disturbance required for soil health plantings. These all in one, food plot type planters make good use of soil moisture compared to other planters that require a prepared seed bed. They perform all the tasks in a single pass reducing the amount of time broken soil is exposed to the air before being resealed by the cultipackers.

An all in one planter with 3 seed box configuration

In the above figure, the front disks prep the seed bed. Seed falls to the ground behind the disks, and the rear ultipackers seal it. Note the paint missing on the fluted disk indicating the shallow soil disturbance of this planter. The angle of attack on the disks is adjustable vie the orange handle to increase or decrease the soil disturbance.

The Till 'n Seed model from Brillion is an all in one planter that creates a very shallow seed bed, less than an inch. The counter rotating sprockets can be seen in the photo under the planter.

Common Problems to Avoid with Planters and Grain Drills

- Planters and seed drills are complex machines that require routine maintenance such as lubrication and cleaning. Store them out of the weather if possible.
- Always clean the seed from the boxes after planting: A large shop vacuum is the easiest method to remove seed. Seed left in the box could become wet and mold or germinate in the seed box creating oxidation and corrosion of the internal mechanisms and clog the drop tubes. Seed also attracts mice and rats which will chew plastic and create potential health hazards.
- Always inspect the planter or drill before use: Spiders, insects, and mice can clog drop tubes with webs and debris when planters sit idle. Use an air compressor to blowout the drop tubes to make sure they are clear. Check to see that all the tubes are dropping seed when calibrating the planter. Check the planter often when in use to insure all seed boxes have adequate, evenly distributed seed, and that the drop tubes are putting out seed.
- Never back up with the planter down, especially planters with "V" openers. This pushes rocks and soil into the openers and drop tubes jamming them. "V" openers that cannot rotate freely will not work properly.
- Visually check the drive mechanism often while in use: This is very important with planters that are driven by disks or coulters. Debris, particularly wood, can get stuck on these and prevent them from rotating or functioning properly. If the drive disk isn't moving the planter is not planting.
- Always check the planting depth by planting a short distance and gently digging down until you find seed. Make adjustments as needed to the planter or the hydraulic setting of the tractor. This is especially critical with no-till drills. The 3rd link arm on the tractor maybe lengthened or shortened to change the angle of the planter and help with depth adjustments or the depth at which the front coulter cuts on a no-till drill.
- On planters with "V" openers, make sure the press wheels are sealing the opening. Make adjustments as needed to insure a firm seal without excessive packing.
- Avoid planting into soft, unpacked seed beds with drill type planters. The loose soil will sluff off into the tracks left by the press wheels and bury seed too deeply after the first rain.

- When planting native or fluffy seed, always use debeared seed if possible. Add a dry seed lubricant such as powdered graphite to facilitate seed flow and reduce bridging. Check drop tubes often during plantings to insure they are putting out seed. A planter with a picker wheel mechanism will greatly increase reliability.
- Use the appropriate seed boxes: If small seed is mixed with large seed in the standard seed box, vibration from use will settle the majority of it at the bottom of the box. This will more than likely result in a higher than desired rate for the small seed, causing the user to run out of small seed, and non-uniform seed distribution.

Seed Drill

A seed drill with mechanical seed metering device mainly consists of : (i) Frame (ii) Seed box (iii) Seed metering mechanism (iv) Furrow openers (iv) Covering device (vi) Transport wheels.

- Frame: The frame is usually made of angle iron with suitable braces and brackets. The frame is strong enough to withstand all types of loads in working condition.
- Seed box: It may be made of mild steel sheet or galvanized iron with a suitable cover. A small agitator is sometimes provided to prevent clogging of seeds.
- Covering device: It is a device to refill a furrow after the seed has been placed in it. Covering the seeds are usually done by patta, chains, drags, packers, rollers or press wheels, designed in various sizes and shapes.
- Transport wheel: There are two wheels fitted on the main axle. Some seed drills have got pneumatic wheels also. The wheels have suitable attachments to transmit power to operate seed dropping mechanism.

Seed Metering Mechanism

The mechanism of a seed drill or fertilizer distributor which delivers seeds or fertilizers from the hopper at selected rates is called seed metering mechanism. Seed metering mechanism may be of several types: (a) Fluted feed type (b) Internal double run type (c) Cup feed type (d) Cell feed mechanism (e) Brush feed mechanism (f) Auger feed mechanism (g) Picker wheel mechanism (h) Star wheel mechanism.

a. Fluted Feed Type: It is a seed metering device with adjustable fluted roller to collect and deliver the seeds into the seed tube. Fluted feed type mechanism consists of a fluted wheel, feed roller, feed cut-off and adjustable gate for different sizes of grains. The feed roller and the feed cut-off device are mounted a shaft, running through the feed cups. The roller carries grooves throughout its periphery. It rotates with the axle over which it is mounted throws the grains out on the adjustable gate from where it falls into the seed tube. The fluted rollers which are mounted at the bottom of the seed box, receive seeds into longitudinal grooves and pass on to the seed tube through the holes provided for this purpose. By shifting the fluted wheel sideways, the length of the grooves exposed to the seed can be increased or decreased and hence the amount of seed is controlled.

b. Internal Double Run Type: It is a seed metering device in which the feed wheel is provided with fine and coarse ribbed flanges. It consists of discs, mounted on a spindle and housed in a casing fitted below the seed box. It has double faced wheel. Internal double-run type roller one face has a larger opening for larger seeds and the other face has smaller opening for smaller seeds. A gate is provided in the bottom-of the box to cover the opening not in use. The rate of seeding is varied by adjusting the speed of the spindle which carries the discs.

c. Cup Feed Mechanism: It is a mechanism consisting of cups or spoons on the periphery of a vertical rotating disc which picks up the seeds from the hopper and delivers them into the seed tubes. It consists of a seed hopper which has two parts. The upper one is called grain box and the lower one is called feed box. The seed delivery mechanism consists of a spindle, carrying a number of discs with a ring of cups attached to the periphery of each disc. The spindle with its frame and attachment is called seed barrel. When the spindle rotates, one disc with its set of cups rotates and picks up few seeds and drops them into small hoppers. The cups have two faces, one for larger seeds and the rate at which the seed barrel revolves. This type of mechanism is common on British seed drills.

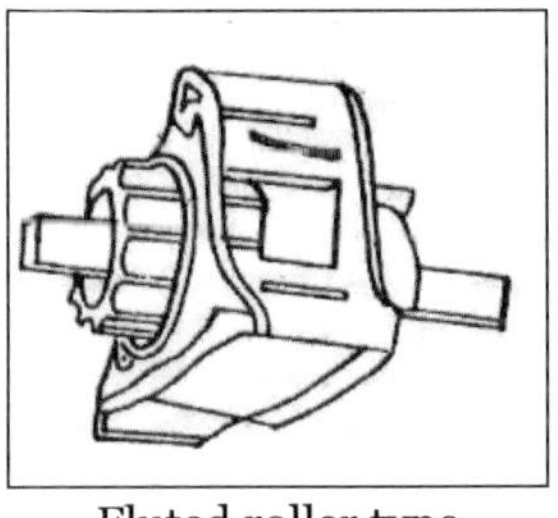

Fluted roller type

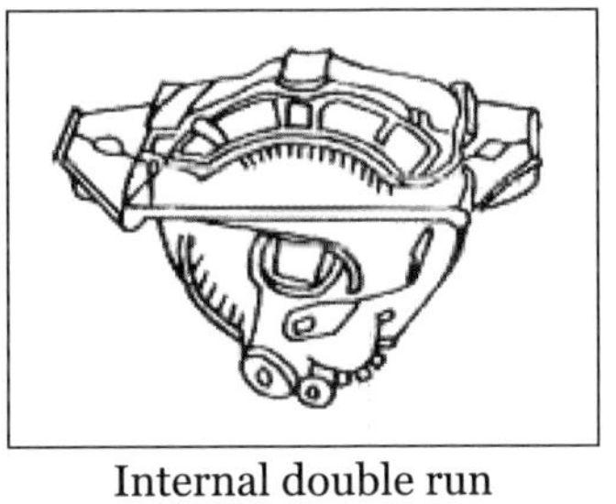

Internal double run

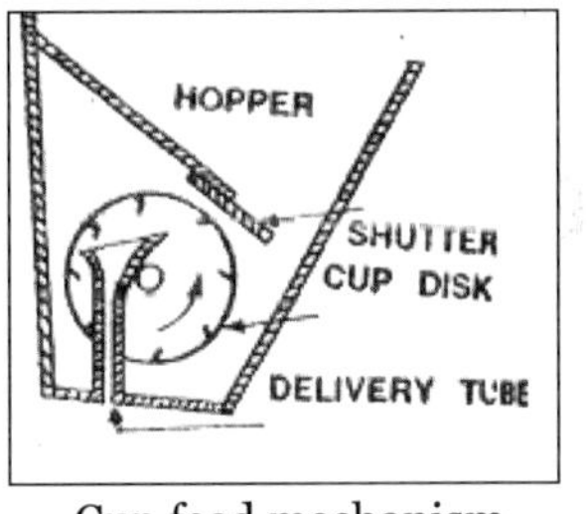

Cup feed mechanism

d. Cell Feed Mechanism: It is a mechanism in which seeds are collected and delivered by a series of equally spaced cells on the periphery of a circular plate or wheel.

e. Brush Feed Mechanism: It is a mechanism in which a rotating brush regulates the flow of seed from the hopper. A number of bullock drawn planters in the country have Brush feed mechanism.

f. Auger Feed Mechanism: It is a distributing mechanism, consisting of an auger which causes a substance to flow evenly in the field, through an aperture at the base or on the side of the hopper. Many of the fertilizer drills on the country have got Auger feed mechanism.

g. Picker Wheel Mechanism: It is a mechanism in which a vertical plate is provided with radially projected arms, which drop the large seeds like potato in furrows with the help of suitable jaws.

h. Star Wheel Mechanism: It is a feed mechanism which consists of a toothed wheel, rotating in a horizontal plane and conveying the fertilizer through a feed gate below the Star wheel.

Furrow Openers

The furrow openers are provided in a seed drill for opening a furrow. The seed tube conducts the seed from the feed mechanism into the boot from where they fall into the furrows.

Type of Furrow Openers

Different type of furrow openers are: (1) Shovel type, (2) Shoe type and (3) Disc Type (single disc, double disc).

1. Shovel Type- Shovel type furrow opener are widely used in seed drills. There are three of shovels in use. They are:

- Reversible shovel,
- Single point shovel and
- Spear point shovel.

Shovel type openers are best suited for stony or root infested fields. These shovels are bolted to the flat iron shanks at the point where boots are fitted which carry the end of the seed tubes. In order to prevent shock loads due to obstructions, springs are provided. It is easy in construction, cheaper and easily repairable. It is very common with usual seed drill.

2. Shoe Type- It works well in trashy soils where the seed beds are not smoothly prepared. They are made from two flat pieces of steel welded together to from a cutting edge. It is specially suited for black cotton soil. Shoe is made of carbon steel having minimum carbon content of 0.5 per cent with a minimum thickness of 4 mm.

3. Disc Type- They are two types: (a) Single disc type and (b) Double disc type.

- Single disc type - It is furrow opener consisting of one concave disc. Disc type furrow openers are found suitable where plant debris or trash mulches are used.
- Double disc type - In double disc type furrow opener there are two flat discs, set at an angle to each other. It is suitable for trashy lands. Seed drills attached with tractors having high speeds, usually use this type of furrow opener.

The furrow opener consists of: (1) tine (2) shovel (3) seed tube (4) boot for seed and fertilizer.

- Shovel - It is made of carbon steel having carbon content of 0.5 per cent and a minimum thickness of 4.0 mm.
- Seed tube - It is a tube which carries the seeds from the metering device to the boot. Seed tubes are provided at the lowest lines through suitable boots and furrow openers. The minimum diameter of seed tube is 25 mm.
- Boot - It is a part of the sowing machine which conveys the seeds or fertilizers from the delivery tube to the furrow. It is bolted or welded to the tine.

Calibration of Grain or Seed Drills

The seed delivery system in drills is not as precise as that used in planters because they use flutes or sponges to meter seed instead of seed singulation.

However, drills are cost effective for establishment of close-spaced crops and are capable of placing seed at a uniform depth, provided depth-control or furrow-closing wheels are properly adjusted and maintained.

Seed charts provided by drill manufacturers give a good starting point for obtaining the desired seeding rate, but they may not be very accurate. Variables that can affect seeding rates include differences in seed size and shape among crop varieties (and sometimes seed lots), seed coatings or seed treatments, and seed density (bushel weight). From time to time, new species or mixtures of two or more species that are not listed on the drill seeding chart will be planted. For these reasons, a producer needs to know how to calibrate a drill. This fact sheet gives easy steps to calibrate your drill. Additionally, it is useful to check actual seeding rates by maintaining records of seed used and acres drilled over the years.

Drill Calibration

1. Determine the area "seeded" when driving the calibration distance: Calibrate the drill for a driving distance of 200 feet. This is an arbitrary length, but it is long enough to capture variability in the field yet not too long to be impractical. Next, determine the width seeded by your drill. In this example we will use a drill with a working width of 10 feet. Working width is calculated by multiplying the spacing between seed openers by the number of openers on the drill. The area seeded in our example will be 2,000 square feet (200 feet long by 10 feet wide). To calculate the fraction of an acre, divide 2,000 square feet by 43,560 square feet per acre to arrive at an area of 0.046 acre.
2. Determine the number of revolutions of the grain drill drive wheel when traveling the calibration distance: In the field use a measuring tape or other measuring device to measure a course that is 200 feet in length. Determine which wheel is responsible for driving the seeding mechanism on the drill. On some drills this is a special wheel that only runs the seed metering mechanism, while on other drills it is one of the wheels that supports the drill. You need to determine how many revolutions this wheel makes in the field in a 200 foot length. Make a mark or apply a piece of heavy tape to the wheel to enable you to easily count wheel revolutions while driving. Drive 200 feet, which you previously marked out in the field, and count the number of revolutions made during the 200 feet. For increased accuracy, repeat this one or two times and use the average.
3. Set up the drill for calibration: Stationary calibration is accomplished by simulating drill operation. The seed delivery system of the drill is operated much in the same way as it is when being pulled across the field by turning the drive wheel the number of turns determined in step 2 above. Stationary calibration is done by lifting the drive wheel above the soil surface so it can freely spin. Some type of hydraulic shop jack is often used to accomplish this. Keep the drill securely attached to the tractor to maintain a level of safety when elevating the drive wheel and to operate the hydraulics on the drill. It is often necessary to lower the drill, permitting the drive mechanism to become engaged so that seeding rate calibration can be accomplished.

Using the rate chart supplied by the drill manufacturer (often found on the underside of the seed box lid), adjust the seed meter for the desired seeding rate of the species of interest. If your seed is not included on the chart, estimate a seeding rate by selecting a similar seed size. This will give you a first approximation. Ensure that enough seed has been added so that seed cups are full and seed metering fluted wheels or sponges are completely covered and will remain covered as seed is metered during the calibration process. Seed can be collected by one of two methods. The first involves the use of a large tarp that can be placed under all drill openers so that seed falling through the seed delivery tubes and disk openers can be collected from below the seed delivery mechanism and weighed. The alternate method involves removing the seed tubes from the disk opener assembly and placing them into several buckets or other containers used to catch and transfer all seed to one container for weighing. At this time, the drive wheel can be turned (in the correct direction) until seed begins falling from all the openers. If some openers do not drop seed, ensure that seed tubes are not kinked or obstructed (straw, spider webs). Now you are set to start collecting seed for calibration. Ensure that the drive wheel has been turned to a convenient starting point (a mark on the drive wheel with chalk, marker, or tape is suggested) and that the tarp or buckets are clean. It is best to collect seed from all openers due to the small area that is represented for this calibration.

4. Calibrate the drill: Turn the drive wheel the number of revolutions needed to represent 200 feet of travel, as determined during step 2 above. Collect all the seed from the tarp or buckets and weigh the total. Use a balance with a gram scale and convert from grams to pounds by dividing by 454 grams per pound. Calibration is a trial-and-error process. The seed is collected from the simulated seeding of a fraction of an acre, weighed, and compared to the desired seeding rate. If the rate is too low, the metering mechanism is adjusted to deliver more seed and the process is repeated. It is desirable to maintain a record of the seeding rates for each setting as one works toward the desired rate(s). Determining the seeding rate of the drill in this example is completed by dividing the total seed weight by the area "drilled" (0.046 acres). For example, 2.3 pounds collected represents a seeding rate of 50 pounds per acre (2.3/0.046 = 50). If you want to check each opener for consistency, collect seed from individual seed tubes and record the weight of the seed from each tube. Although the seed delivery system on a drill is not as precise as the seed-metering system on most corn planters, an amount collected from one opener that is 10 percent different from the average on all openers indicates a need for maintenance to the seed-metering mechanism.

Considerations in Determining Seeding Rate and Seed Purchase

A number of considerations are pertinent when determining seeding rates. A good source for seeding rate recommendations is your state's agricultural Extension office. Seeding rates are often recommended by seed suppliers for their specific species and varieties. These rates may be modified by planting date, soil type, whether drilling into tilled or no-tilled conditions, soil fertility levels, and other factors. Generally recommended rates are based on the desired plant population following emergence. Seeding rates in pounds per acre will be determined with the knowledge of how many seeds are in one pound of seed. Once this information is known, the seeding rate should be modified to account for percent germination and the amount of non-seed that is indicated on the seed label. Hard seed is defined as seed that will not germinate as quickly as the other pure seed in the bag. Hard seed is common in legumes, but seed rates usually do not need to be adjusted for hard seed. Check with your seed supplier if you have seed with more than 5 percent hard seed. When selecting seed, always review the seed labels for purity and germination. Pure live seed (PLS) is required by law to be stated on labels for all seed sold in Pennsylvania. PLS = purity x germination, or the percentage of the bulk seed weight that is viable and should germinate when planted. Keep in mind that PLS may vary among seed lots from the same supplier.

Calibration for Mixtures

Seed mixtures are becoming more popular with increased understanding of the complementarity of different species grown for forage or as a cover crop. Calibrating the drill for mixtures is no different from calibration of the drill for single species. As a first approximation, determine the seeding rate of each species in the mixture and divide the normal seeding rate of each species by the number of species mixed. So if, for example, you mix hairy vetch and rye and the normal seeding rate is 20 pounds per acre for hairy vetch and 2 bushels per acre for rye, you may plant 10 pounds per acre of hairy vetch and 1 bushel per acre of rye. To compose the mixture, determine how many acres to plant, weigh the amount of seed of each species, and mix thoroughly. If a two-species mixture is planted and each species can be seeded through a different box, you can calibrate each box for that species individually. If the seeds are mixed prior to calibration, calibrate using the guidelines for single species. Finding the first setting may be the most challenging. In many cases, mixtures will pack more densely, so start with the setting for seeding the species with the largest seed and then adjust accordingly. In the above example, you might start with the setting for 1 bushel per acre of rye.

Fine Tuning the Calibration

After calibrating the drill, you can keep track of the seeding rate after drilling a few acres. True seeding rate can be determined by one of several methods.

Most modern grain drills are equipped with an acre meter. Use it to monitor seeding rate. Keep track of the seed you have used and divide the amount of seed drilled by the acres planted. Alternatively, if an acre meter is not available on the drill and acreage is known for a given field, calculate the rate in a similar fashion. If a minor adjustment is needed, use the seed rate increments between settings to more accurately make the adjustment. Actual seeding rates may vary somewhat due to groundcover, soil conditions, overlapping between passes, and driving speed.

Drill Adjustment to Ensure Consistent Plant Establishment

Depth should be checked several times, especially when beginning the planting operation until proper depth is achieved, and then periodically during seeding. When operating in no-till conditions, depth adjustments may need to be made as amounts of residues from previous crops change across the field. Be careful not to plant small-seeded species--particularly small-seeded legumes--too deep, especially during the spring, in cold soils. Seeding depth should be roughly 3.5 times seed diameter. Use no more down pressure on the drill and closing wheels than necessary to provide consistent seed-to-soil contact. Too much down pressure, especially in wet soils, provides a very poor environment for emerging seedlings. Pay particular attention to the drive wheel, if it is not also a transport wheel, to ensure that it continuously rides on the soil surface to ensure consistent seed metering.

Drill Performance

Many drills with a small seed box have tubes that drop the seed behind the double disk openers, just before the press wheel. In some cases this may mean a significant portion of seed is not in close contact with the soil. Some farmers have found it beneficial to run a cultipacker through the field after drilling to increase seed-to-soil contact. Drill wear also needs to be taken into consideration when calibrating the drill. When changing the opening setting on older drills, not all openings may be affected in the same manner. If this becomes extreme, the drill may be in need of maintenance.

Seed Processing Equipments

Step 1: Pre Conditioning and Pre Cleaning

Pre conditioning: Isolation of seed from plant parts with which it was harvested e.g. Shelling.

Pre Cleaning

Removal of external materials like trash, stones, clods which are either in larger size or lighter in weight. No pre cleaning is required for hand harvested and winnowed seeds.

The machineries involved in this operation are:

1. Scalper: It removes the larger inert matter from the seeds. If it contains a single sieve it is called as scalpers, two sieves – rough cleaners. The unit consists of a vibrating or rotating screen or sieve having perforation large enough to allow the rough seed pass through readily.

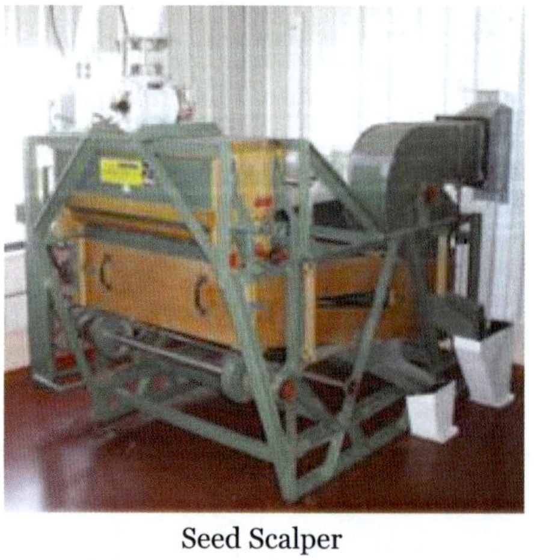

Seed Scalper

2. Debearders : The machine has horizontal beater with arms rotating inside a steel drum. When the seeds pass through it do the action of rubbing the seeds and clip the seeds of oats, debeard barley, thresh white cap in wheat, remove awns and beards, de hull some grass seeds and polish the seed.

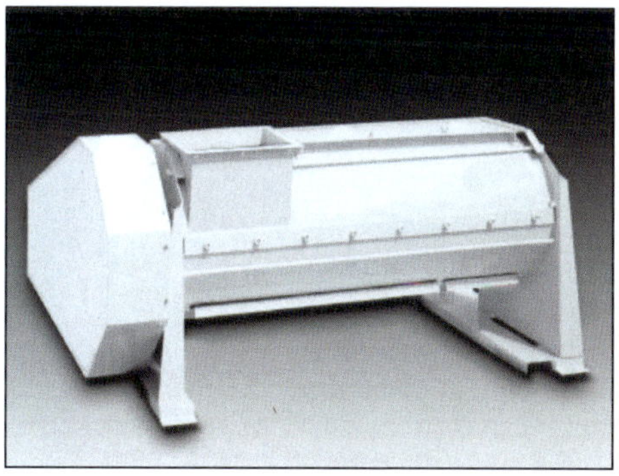

Debearders

Huller Scarifier

Have two rubber faced rough surfaces to rub the seeds. Dehulling (removal of outer coat or husk) and scarifier (scratching the seed coat) can be done simultaneously or separately. Its operations, Seed, rotating disc, centrifugal force, thrown - huller, suction chamber removes lighter seeds.

Maize sheller:

1. High capacity power operated shellers – bulk.
2. Hand shellers – breeder or nucleus seed.

Step 2: Cleaning

The second stage of cleaning is carried out with air blasts and vibrating screens and is applicable to all kinds of seeds. It is essentially the same as scalping but more refined. It is performed mostly by one machine known as air-screen cleaner.

Air-screen Cleaner Cum Grader

The air-screen machine is the basic cleaner in most seed processing plants. Almost all seed must be cleaned by air-screen cleaner before specific specifications can be attempted. Machine size varies from small, two-screen farm models to large industrial cleaners with 7-8 screens. Two-screen models are used on farms, in breeder and foundation seed programs and by experiment stations for processing small quantities of seed. In most machines separations are made on the basis of differences in only one physical characteristic.

The air-screen machine, however, effects separations on the basis of differences in size and weight of seeds. This enables the air-screen machine to use three cleaning elements: aspirator, in which light material is removed from the seed mass; scalper in which good seed are dropped through screen openings; but larger material is carried over the screen into a separate spout; and grader, in which good crop seed ride over screen openings, while smaller particles drop through.

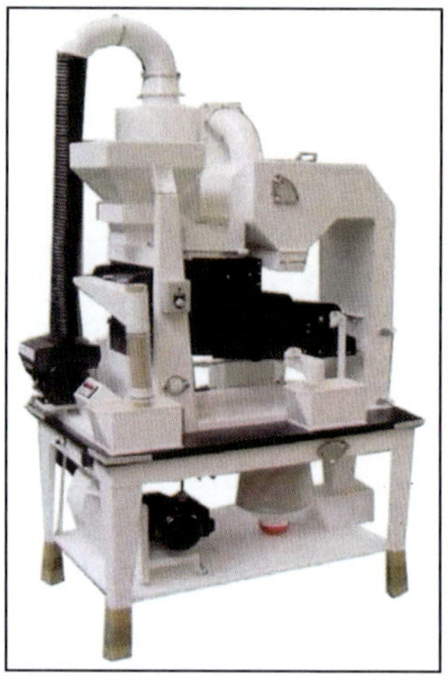

Step 3: Cleaning and Grading

To obtain quality seed, it is necessary to clean the seed obtained from the farm to get rid of inert materials, weed seeds, other crop seeds, other variety seeds, damaged and deteriorated seed. Different kinds of seeds can be separated when they differ in one or more physical characteristics. Physical characteristics normally used to separate seeds are size, shape, length, weight, colour, surface texture, affinity to liquids, electrical conductivity, etc. The problem lies in identifying the most important property and use the machine that separates seed using the identified property. Some of the identified properties and machines operating by following the properties are listed in the table below:

Name of the Separator	Property followed	Uses
Vibratory separator	Shape and surface texture	Removal of weed seeds
Spiral separator	Shape or the degree of its ability to roll	Separation of damaged/flat and wrinkled seeds from smooth seeds. Separation of mustard, rape, soybean and peas from wheat, flax, oats, etc., and round seeds from flat seeds.
Disk/Indented cylinder separator	Length	Dissimilar material like wheat, rye, mustard, barley from oats
Electrostatic separator	Electrical property	Johnson grass from sesame seed
Electronic colour sorters	Colour/brightness	Separation of off coloured seeds
Inclined draper	Shape and surface texture	Separation of smooth or round seeds from rough flat or elongated seeds
Magnetic separator	Surface texture and stickiness	Removal of contaminating weed seed from clovers, alfalfa seeds and iron metals
Roll mill	Shape and surface texture	Separation of smooth clover seed
Gravity separator or De-stoner	Density or specific gravity	Removal of badly damaged, deteriorated, insect damaged crop seed and stones from good seeds.

Upgrading

Seed lots require further cleaning treatment to remove adulterants that are similar to pure seed in size and shape, to be separated by air screen cleaner. Removal of seeds larger or smaller than required size (sizing) and removal of cracked, damaged or otherwise defective seeds (grading) is accomplished in this final stage of processing.

Specific Gravity Separation

This method makes use of a combination of weight and surface characteristics of the seed to be separated. The principle of floatation is employed here. A mixture of seeds is fed onto the lower end of a sloping perforated table. Air is forced up through the porous deck surface and the bed of seeds by a fan, which stratifies the seeds in layers according to density with the lightest seeds and particles of inert matter at the top and the heaviest at the bottom. An oscillating movement of the table causes the seeds to move at different rates across the deck. The lightest seeds float down under gravity and are discharged at the lower end, while the heaviest ones are kicked up the slope by contact with the oscillating deck and are discharged at the upper end. This machine separates seeds of the same density but of different size and seeds of the same size but of different densities.

Specific gravity separator

Indented Cylinder

This helps to separate seeds according to the length. The equipment consists of a slightly inclined horizontal rotating cylinder and a movable separating trough. The inside surface has small closely spaced hemispherical indentations. Small seeds are pressed into the indents by centrifugal force and can be removed. The larger seeds flows in the centre of the cylinder and is discharged by gravity.

Magnetic Separator

The magnetic separator separates seed according to its surface texture or related seed characteristics. First, seed is treated with iron filings, which adhere to rough surface alone. The treated seed lot is passed over a revolving magnetic drum and separated from smooth, uncoated seed. It may help to add varied amounts of water while mixing seed and powder, depending on the seed type. At any rate, the effectiveness of magnetic separation depends on the components of the seed lot and on the powder and water used in the treating operation. The greater the difference between surface textures of the seed lot's components, more effective will be the separation.

Principle of Function

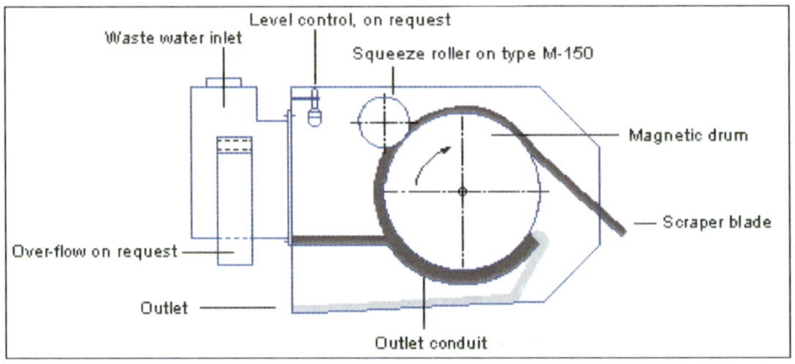

Colour Separator

The colour separator is used to separate discoloured seed, greatly of lower quality. Separation based on colour is necessary because the density and dimensions of discoloured seed are the same as those of sound seed, so other machines are not effective for separation. Electronic colour separation uses photocells to compare the seed colour with "background" which are selected to reflect the same light as the good seed. Seed that differs in colour is detected by the photo cells, which generate an electric impulse. The impulse activates an air jet to blow away the discoloured seed.

Separating Seed by Color

The color sorter (right) uses an electronic eye that can pick up different colors according to the way the machine is adjusted. As seed falls down a shoot, it passes through the electric eye. If the color of the seed is different than the desired color, the electric eye will activate a sudden burst of air that pushes that seed into a reject bin while the rest of the seed passes through to another bin.

Friction Cleaning

The air-screen combinations cannot remove debris that has a size and density similar to the seeds. However, if the debris has a different surface texture, it may be possible to remove by friction cleaning. Any object rolling or sliding over a sloping surface encounters a certain friction depending on the texture of itself and that of the sloping surface. Separation is made on a velvet cloth or rubber belt with variable inclination, which ensures that the slope necessary for the run off of the seed is different from the slope necessary for run -off of the debris. The belt continuously moves upwards and removes the debris while the seeds roll down the slope.

Spiral Separator

The separator, which classifies seed according to its shape and rolling ability, consists of sheet metal strips fitted around a central axis in the form of a spiral. The unit resembles an open screw conveyor standing in a vertical position. The seed is introduced at the top of the inner spiral. Round seeds roll faster down the incline than flat or irregularly shaped seeds, which tend to slide or tumble. The orbit of round seed increases with speed on its flight around the axis, until it rolls over the edge of the inner flight into the outer flight where it is collected separately. The slower moving seed does not build up enough speed to escape form the inner flight. Most spirals have multiple inner flights arranged one above the other to increase the capacity.

Liquid Flotation

Cleaning by flotation relies on the principle that the density of the seed of a given species is specific both for filled and ill filled seed. In this method, liquids with a density or specific gravity between that of the full and empty seed are used. The specific gravity of the liquids used is such that the full seed sinks and the empty seed and light debris float.

Disc Harrows

Disc harrows are farming implements used to prepare the soil for planting or sowing by breaking up the clods and surface crusts, thus improving soil granulation and destroying the weeds. They consist of concave cutting blades (disks), mounted on a common shaft, that form a gang. The disc harrow consists of the following parts:

1. Frame: Holds and supports the disc harrow.

2. Disc: Circular concave cutting blade made of steel. There are the two types of discs:
   - Smooth edge disc; also called plain edge disc; consists of plain discs used for normal soil conditions.
   - Notched edge disc, also called cut-away discs or scalloped edge discs with serrated edges; used for cutting crop residues and weeds.

Smooth edge disc (left) and notch edge disc (right)

3. Arbor bolt: Also called gang bolt, is a long heavy steel shaft on which discs are mounted.
4. Spool or spacer: Component mounted on arbor bolet between every two discs to retain their fixed position and prevent any lateral movement of the disc.
5. Bearings: Essential for providing rotation of the gang and regulating the thurst.
6. Bumper: Heavy iron plate situated at the end of each gang to counter; protects the discs from collision of adjacent discs.
7. Scraper: Removes the soil from the disks, keeping the concave side of the disk clean, thus preventing the clogging of the discs.
8. Disc gang: Each group of disks mounted on a common arbor bolt with disc spacer, bearing, and bumper; may consist of 3 to 13 discs.
9. Weight box: Included on the frame to provide additional weight for increasing the penetration of the disc into the soil.

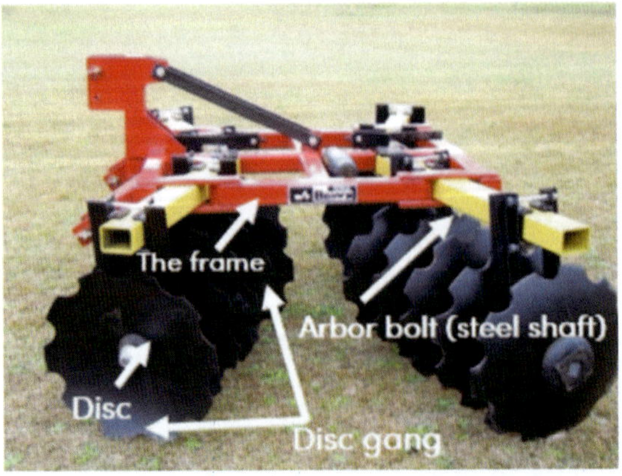

Important parts of disc harrow

Types of Disc Harrows

There are various types of disc harrows. Regarding of the operation mode, there are three different types of disc harrows:

- Single Action Disc Harrow: Consists of two disc gangs arranged in opposite directions, it throws the soil in opposite directions creating ridges and furrows.
- Double Action Disc Harrow: Also called tandem disc harrow; consists of two or more gangs, discs from the front gangs throw the soil in one direction, while discs on the rear gangs throw the soil in the opposite direction.

Single action disc harrow (left) and double action disc harrow (right)

- Offset Disc Harrow: Consists of two gangs (left and right) operating in tandem; the harrow is usually placed in the offside of the tractor so it is not in the same line of pulling the tractor.

Off-set disc harrow

Aside from the aforementioned classifications, disc harrows can be classified according to the type of mounting used with the tractor (mounted, semi-mounted, trailed), as well as by the size of the disc.

In regards to the disc diameter, there are three different disc harrows:

- Light disc harrows; with a disc diameter of 20-30 cm.
- Middle disc harrows, with a disc diameter of 30-50 cm.
- Heavy disc harrows, with a disc diameter more than 60 cm.

Notched edge disks are used for entering the crop residue

The type of harrow disk a farmer chooses for soil preparation management will depend on the purpose of disking, whether it's intended for entering the crop residue or simply leveling the soil structure. Another important aspect to consider when choosing the most appropriate type of disc harrow is the size of the field.

Larger fields demand improved farm equipment, while smaller fields can be managed with the simple farm implements such as the single action disc harrow. So make sure you choose the most appropriate disc harrows for your soil preparation management.

Agriculture and Biofuels

Biofuels are fuels developed by converting biomass into liquid fuels for transportation. Sources of biomass for biofuels include:

- Grassy and woody plants,
- Agriculture or forestry residues,
- Algae,
- Crops such as corn and soybeans,
- Vegetable oils and animal fats.

Biofuels have the potential to reduce greenhouse gas emissions and reduce our dependence on foreign oil. The two most common types of biofuels in use today are ethanol and biodiesel.

Traditional biomass, including fuelwood, charcoal and animal dung, continues toprovide important sources of energy in many parts of the world. Bioenergy is the dominant energy source for most of the world's population who live in extreme poverty and who use this energy mainly for cooking. More advanced and efficient conversion technologies now allow the extraction of biofuels – in solid, liquid and gaseous forms – from materials such as wood, crops and waste material. This chapter provides an overview of biofuels. What are they, what is their potential and what are their implications for agriculture? The main focus, however, is on liquid biofuels for transport, which are now gaining in prominence as a result of the rapid increase in their use.

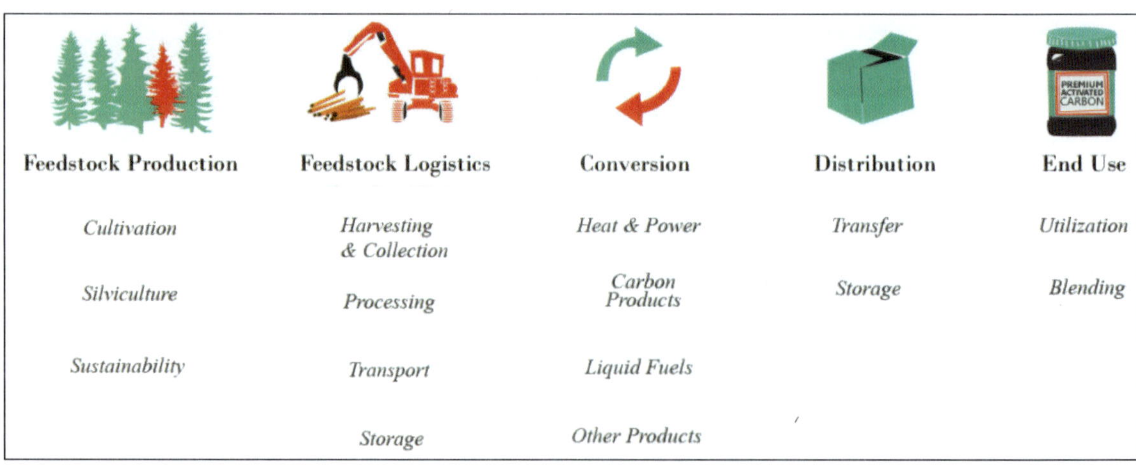

Biofuels – from feedstock to end use

Types of Biofuels

Biofuels are energy carriers that store the energy derived from biomass. A wide range of biomass sources can be used to produce bioenergy in a variety of forms. For example, food, fibre and wood process residues from the industrial sector; energy crops, shortrotation crops and agricultural wastes from the agriculture sector; and residues from the forestry sector can all be used to generate electricity, heat, combined heat and power, and other forms of bioenergy. Biofuels may be referred to as renewable energy because they are a form of transformed solar energy.

Biofuels can be classified according to source and type. They may be derived from forest, agricultural or fishery products or municipal wastes, as well as from agroindustry, food industry and food service by-products and wastes. They may be solid, such as fuelwood, charcoal and wood pellets; liquid, such as ethanol, biodiesel and pyrolysis oils; or gaseous, such as biogas.

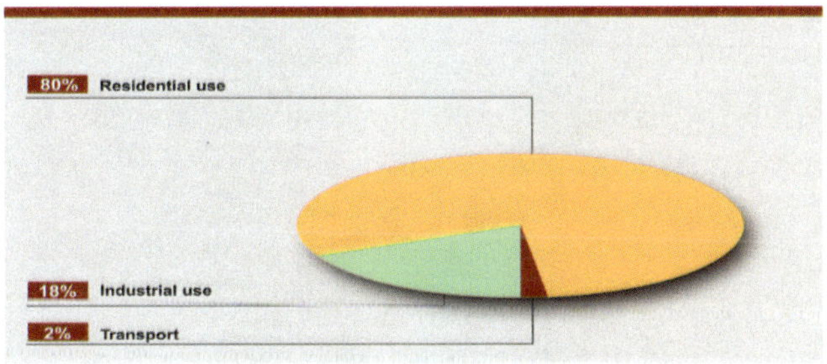

Uses of biomass for energy

A basic distinction is also made between primary (unprocessed) and secondary (processed) biofuels:

- Primary biofuels, such as firewood, wood chips and pellets, are those where the organic material is used essentially in its natural form (as harvested). Such fuels are directly combusted, usually to supply cooking fuel, heating or electricity production needs in small and largescale industrial applications.
- Secondary biofuels in the form of solids (e.g. charcoal), liquids (e.g. ethanol, biodiesel and bio-oil), or gases (e.g. biogas, synthesis gas and hydrogen) can be used for a wider range of applications, including transport and high-temperature industrial processes.

Liquid Biofuels for Transport

In spite of their limited overall volume, the strongest growth in recent years has been in liquid biofuels for transport, mostly produced using agricultural and food commodities as feedstocks. The most significant are ethanol and biodiesel.

Ethanol

Any feedstock containing significant amounts of sugar, or materials that can be converted into sugar such as starch or cellulose, can be used to produce ethanol. Ethanol available in the biofuel market today is based on either sugar or starch. Common sugar crops used as feedstocks are sugar cane, sugar beet and, to a lesser extent, sweet sorghum. Common starchy feedstocks include maize, wheat and cassava. The use of biomass containing sugars that can be fermented directly to ethanol is the simplest way of producing ethanol. In Brazil and other tropical countries currently producing ethanol, sugar cane is the most widely used feedstock. In OECD countries, most ethanol is produced from the starchy component of cereals (although sugar beet is also used), which can be converted fairly easily into sugar. However, these starchy products represent only a small percentage of the total plant mass. Most plant matter is composed of cellulose, hemicellulose and lignin; the first two can be converted into alcohol after they have first been converted into sugar, but the process is more difficult than the one for starch. Today, there is virtually no commercial production of ethanol from cellulosic biomass, but substantial research continues in this area.

Ethanol can be blended with petrol or burned in its pure form in slightly modified spark-ignition engines. A litre of ethanol contains approximately 66 percent of the energy provided by a litre of petrol, but has a higher octane level and when mixed with petrol for transportation it improves the performance of the latter. It also improves fuel combustion in vehicles, thereby reducing the emission of carbon monoxide, unburned hydrocarbons and carcinogens. However, the combustion of ethanol also causes a heightened reaction with nitrogen in the atmosphere, which can result in a marginal increase in nitrogen oxide gases. In comparison with petrol, ethanol contains only a trace amount of sulphur. Mixing ethanol with petrol, therefore, helps to reduce the fuel's sulphur content and thereby lowers the emissions of sulphur oxide, a component of acid rain and a carcinogen.

Biodiesel

Biodiesel is produced by combining vegetable oil or animal fat with an alcohol and a catalyst through a chemical process known as transesterification. Oil for biodiesel production can be extracted from almost any oilseed crop; globally, the most popular sources are rapeseed in Europe and soybean in Brazil and the United States of America. In tropical and subtropical countries, biodiesel is produced from palm, coconut and jatropha oils. Small amounts of animal fat, from fish- and animal-processing operations, are also used for biodiesel production. The production process typically yields additional by-products such as crushed bean "cake" (an animal feed) and glycerine. Because biodiesel can be based on a wide range of oils, the resulting fuels can display a greater variety of physical properties, such as viscosity and combustibility, than ethanol.

Biodiesel can be blended with traditional diesel fuel or burned in its pure form in compression ignition engines. Its energy content is 88–95 percent of that of diesel, but it improves the lubricity of diesel and raises the cetane value, making the fuel economy of both generally comparable. The higher oxygen content of biodiesel aids in the completion of fuel combustion, reducing emissions of particulate air pollutants, carbon monoxide and hydrocarbons.

As with ethanol, diesel also contains only a negligible amount of sulphur, thus reducing sulphur oxide emissions from vehicles.

Straight Vegetable Oil

Straight vegetable oil (SVO) is a potential fuel for diesel engines that can be produced from a variety of sources, including oilseed crops such as rapeseed, sunflower, soybean and palm. Used cooking oil from restaurants and animal fat from meat-processing industries can also be used as fuel for diesel vehicles.

Biofuel Feedstocks

There are many supply sources of biomass for energy purposes, scattered across large and diverse geographical areas. Even today, most energy derived from biomass used as fuel originates from by-products or co-products of food, fodder and fibre production. For instance, the main byproducts of forest industries are used to produce fuelwood and charcoal, and black liquor (a by-product of pulp mills) is a major fuel source for bioelectricity generation in countries such as Brazil, Canada, Finland, Sweden and the United States of America. A considerable amount of heat and power is derived from recovered and recycled woody biomass and increasing amounts of energy are recovered from biomass derived from cropland (straw and cotton stalks) and forest land (wood chips and pellets). In sugar and coffee-producing countries, bagasse and coffee husks are used for direct combustion and to produce heat energy and steam.

In terms of bioenergy, however, the big growth area in recent years has been in the production of liquid biofuels for transport using agricultural crops as feedstocks. The bulk of this has taken the form of ethanol, based on either sugar crops or starchy crops, or biodiesel based on oil crops.

As shown in figure below, a range of different crops can be used as feedstock for ethanol and biodiesel production. However, most global ethanol production is derived from sugar cane or maize; in Brazil, the bulk of ethanol is produced from sugar cane and in the United States of America from maize.

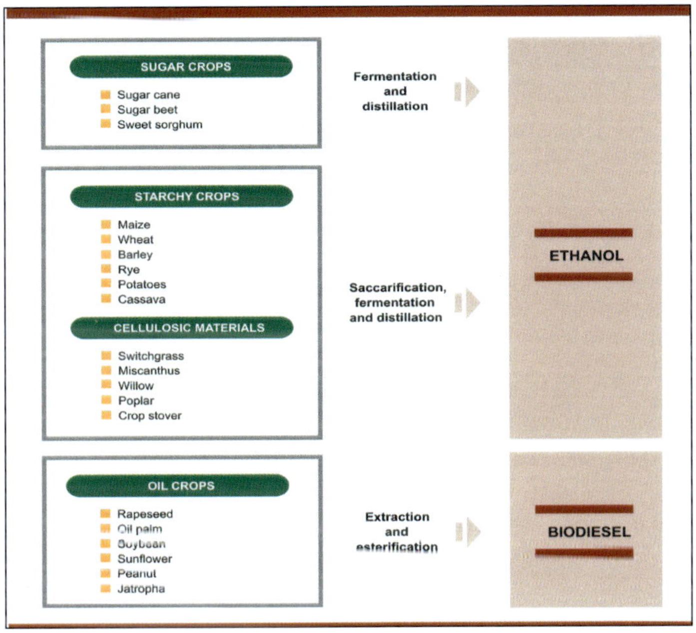

Conversion of agricultural feedstocks into liquid biofuels

Other significant crops include cassava, rice, sugar beet and wheat. For biodiesel, the most popular feedstocks are rapeseed in the EU, soybean in the United States of America and Brazil, and palm, coconut and castor oils in tropical and subtropical countries, with a growing interest in jatropha.

Biofuels and Agriculture

The current expansion and growth of energy markets, as a result of new energy and environment policies enacted over the past decade in most developed countries and in several developing countries, is reshaping the role of agriculture. Most significant is the sector's increasing role as a provider of feedstock for the production of liquid biofuels for transport – ethanol and biodiesel. Modern bioenergy represents a new source of demand for farmers' products. It thus holds promise for the creation of income and employment. At the same time, it generates increasing competition for natural resources, notably land and water, especially in the short run, although yield increases may mitigate such competition in the longer run. Competition for land becomes an issue especially when some of the crops (e.g. maize, oil palm and soybean) that are currently cultivated for food and feed are redirected towards the production of biofuels, or when food-oriented agricultural land is converted to biofuel production.

Currently, around 85 percent of the global production of liquid biofuels is in the form of ethanol. The two largest ethanol producers, Brazil and the United States of America, account for almost 90 percent of total production, with the remainder accounted for mostly by Canada, China, the EU (mainly France and Germany) and India. Biodiesel production is principally concentrated in the EU (with around 60 percent of the total), with a significantly smaller contribution coming from the United States of America. In Brazil, biodiesel production is a more recent phenomenon and production volume remains limited. Other significant biodiesel producers include China, India, Indonesia and Malaysia.

Different crops vary widely in terms of biofuel yield per hectare, both across feedstocks and across countries and production systems, as illustrated in table. Variations are due both to differences in crop yields per hectare across crops and countries and to differences in conversion efficiency across crops. This implies vastly different land requirements for increased biofuel production depending on the crop and location. Currently, ethanol production from sugar cane and sugar beet has the highest yields, with sugar-cane-based production in Brazil topping the list of in terms of biofuel output per hectare and India not far behind. Yields per hectare are somewhat lower for maize, but with marked differences between yields, for example, in China and in the United States of America. The data reported in table refer only to technical yields. The cost of producing biofuels based on different crops in different countries may show very different patterns.

The Biofuels Life Cycle: Energy Balances and Greenhouse Gas Emissions

Two of the main driving forces behind policies promoting biofuel development have been concerns over energy security and a desire to reduce greenhouse gas emissions. Just as different crops have different yields in terms of biofuel per hectare, wide variations also occur in terms of energy balance and greenhouse gas emission reductions across feedstocks, locations and technologies.

The contribution of a biofuel to energy supply depends both on the energy content of the biofuel and on the energy going into its production. The latter includes the energy required to cultivate and harvest the feedstock, to process the feedstock into biofuel and to transport the feedstock and the resulting biofuel at the various phases of its production and distribution. The fossil energy balance expresses the ratio of energy contained in the biofuel relative to the fossil energy used in its production. A fossil energy balance of 1.0 means that it requires as much energy to produce a litre of biofuel as it contains; in other words, the biofuel provides no net energy gain or loss. A fossil fuel energy balance of 2.0 means that a litre of biofuel contains twice the amount of energy as that required in its production. Problems in assessing energy balances accurately derive from the difficulty of clearly defining the system boundary for the analysis.

Biofuel production by country, 2007

COUNTRY/COUNTRY GROUPING	ETHANOL		BIODIESEL		TOTAL	
	(Million litres)	(Mtoe)	(Million litres)	(Mtoe)	(Million litres)	(Mtoe)
Brazil	19 000	10.44	227	0.17	19 227	10.60
Canada	1 000	0.55	97	0.07	1 097	0.62
China	1 840	1.01	114	0.08	1 954	1.09
India	400	0.22	45	0.03	445	0.25
Indonesia	0	0.00	409	0.30	409	0.30
Malaysia	0	0.00	330	0.24	330	0.24
United States of America	26 500	14.55	1 688	1.25	28 188	15.80
European Union	2 253	1.24	6 109	4.52	8 361	5.76
Others	1 017	0.56	1 186	0.88	2 203	1.44
World	**52 009**	**28.57**	**10 204**	**7.56**	**62 213**	**36.12**

Biofuel yields for different feedstocks and countries

CROP	GLOBAL/NATIONAL ESTIMATES	BIOFUEL	CROP YIELD	CONVERSION EFFICIENCY	BIOFUEL YIELD
			(Tonnes/ha)	(Litres/tonne)	(Litres/ha)
Sugar beet	Global	Ethanol	46.0	110	5 060
Sugar cane	Global	Ethanol	65.0	70	4 550
Cassava	Global	Ethanol	12.0	180	2 070
Maize	Global	Ethanol	4.9	400	1 960
Rice	Global	Ethanol	4.2	430	1 806
Wheat	Global	Ethanol	2.8	340	952
Sorghum	Global	Ethanol	1.3	380	494
Sugar cane	Brazil	Ethanol	73.5	74.5	5 476
Sugar cane	India	Ethanol	60.7	74.5	4 522
Oil palm	Malaysia	Biodiesel	20.6	230	4 736
Oil palm	Indonesia	Biodiesel	17.8	230	4 092
Maize	United States of America	Ethanol	9.4	399	3 751
Maize	China	Ethanol	5.0	399	1 995
Cassava	Brazil	Ethanol	13.6	137	1 863
Cassava	Nigeria	Ethanol	10.8	137	1 480
Soybean	United States of America	Biodiesel	2.7	205	552
Soybean	Brazil	Biodiesel	2.4	205	491

The above figure summarizes the results of several studies on fossil energy balances for different types of fuel, as reported by the Worldwatch Institute. The figure reveals wide variations in the estimated fossil energy balances across feedstocks and fuels and, sometimes, for a feedstock/fuel combination, depending on factors such as feedstock productivity, agricultural practices and conversion technologies.

Conventional petrol and diesel have fossil energy balances of around 0.8–0.9, because some energy is consumed in refining crude oil into usable fuel and transporting it to markets. If a biofuel has a fossil energy balance exceeding these numbers, it contributes to reducing dependence on fossil fuels. All biofuels appear to make a positive contribution in this regard, albeit to widely varying degrees. Estimated fossil fuel balances for biodiesel range from around 1 to 4 for rapeseed and soybean feedstocks. Estimated balances for palm oil are higher, around 9, because other oilseeds must be crushed before the oil can be extracted, an additional processing step that requires energy. For crop-based ethanol, the estimated balances range from less than 2 for maize to around 2–8 for sugar cane. The favourable fossil energy balance of sugar-cane-based ethanol, as produced in Brazil, depends not only on feedstock productivity, but also on the fact that it is processed using biomass residues from the sugar cane (bagasse) as energy input. The range of estimated fossil fuel balances for cellulosic feedstocks is even wider, reflecting the uncertainty regarding this technology and the diversity of potential feedstocks and production systems.

Similarly, the net effect of biofuels on greenhouse gas emissions may differ widely.

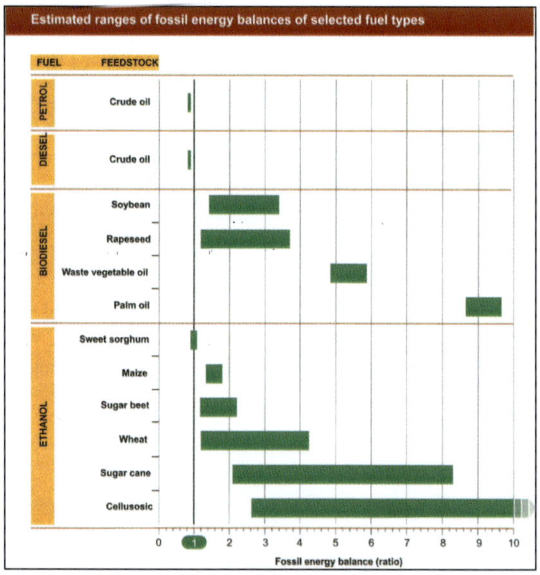

Estimated ranges of fossil energy balances of selected fuel types

Biofuels are produced from biomass; in theory, therefore, they should be carbon neutral, as their combustion only returns to the atmosphere the carbon that was sequestered from the atmosphere by the plant during its growth unlike fossil fuels, which release carbon that has been stored for millions of years under the surface of the earth. However, assessing the net effect of a biofuel on greenhouse gas emissions requires analysis of emissions throughout the life cycle of the biofuel: planting and harvesting the crop; processing the feedstock into biofuel; transporting the feedstock and the final fuel; and storing, distributing and retailing the biofuel including the impacts of fuelling a vehicle and the emissions caused by combustion. In addition, any possible co-products that may reduce emissions need to be considered. Clearly, therefore, fossil energy balances are only one of several determinants of the emissions impact of biofuels. Critical factors related to the agricultural production process include fertilizing, pesticide use, irrigation technology and soil treatment. Land-use changes associated with expanded biofuel production can have a major impact. For example, converting forest land to the production of biofuel crops or agricultural crops displaced by biofuel feedstocks elsewhere can release large quantities of carbon that would take years to recover through the emission reductions achieved by substituting biofuels for fossil fuels.

Second-generation Liquid Biofuels

Current liquid biofuel production based on sugar and starch crops (for ethanol) and oilseed crops (for biodiesel) is generally referred to as first-generation biofuels. A second generation of technologies under development may also make it possible to use lignocellulosic biomass. Cellulosic biomass is more resistant to being broken down than starch, sugar and oils. The difficulty of converting it into liquid fuels makes the conversion technology more expensive, although the cost of the cellulosic feedstock itself is lower than for current, first-generation feedstocks. Conversion of cellulose to ethanol involves two steps: the cellulose and hemicellulose components of the biomass are first broken down into sugars, which are then fermented to obtain ethanol. The first step is technically challenging, although research continues on developing efficient and cost-effective ways of carrying out the process. The lack of commercial viability has so far inhibited significant production of cellulose-based second-generation biofuels.

As cellulosic biomass is the most abundant biological material on earth, the successful development of commercially viable second-generation cellulose-based biofuels could significantly expand the volume and variety of feedstocks that can be used for production. Cellulosic wastes, including waste products from agriculture (straw, stalks, leaves) and forestry, wastes generated from processing (nut shells, sugarcane bagasse, sawdust) and organic parts of municipal waste, could all be potential sources. However, it is also important to consider the crucial role that decomposing biomass plays in maintaining soil fertility and texture; excessive withdrawals for bioenergy use could have negative effects.

Dedicated cellulosic energy crops hold promise as a source of feedstock for secondgeneration technologies. Potential crops include short-rotation woody crops such as willow, hybrid poplars and eucalyptus or grassy species such as miscanthus, switchgrass and reed canary grass. These crops present major advantages over firstgeneration crops in terms of environmental sustainability. Compared with conventional starch and oilseed crops, they can produce more biomass per hectare of land because the entire crop is available as feedstock for conversion to fuel. Furthermore, some fastgrowing perennials such as short-rotation woody crops and tall grasses can sometimes grow on poor, degraded soils where foodcrop production is not optimal because of erosion or other limitations. Both these factors may reduce competition for land with food and feed production. On the downside, several of these species are considered invasive or potentially invasive and may have negative impacts on water resources, biodiversity and agriculture.

Second-generation feedstocks and biofuels could also offer advantages interms of reducing greenhouse gas emissions. Most studies project that future, advanced fuels from perennial crops and woody and agricultural residues could dramatically reduce life-cycle greenhouse gas emissions relative to petroleum fuels and firstgeneration biofuels. This stems from both the higher energy yields per hectare and the different choice of fuel used in the conversion process. In the current production process for ethanol, the energy used in processing is almost universally supplied by fossil fuels (with the exception of sugarcane-based ethanol in Brazil, where most of the energy for conversion is provided by sugar-cane bagasse). For second-generation biofuels, process energy could be provided by left-over parts of the plants (mainly lignin).

While cellulosic biomass is harder to break down for conversion to liquid fuels, it is also more robust for handling, thus helping to reduce its handling costs and maintain its quality compared with food crops. It is also easier to store, especially in comparison with sugar-based crops, as it resists deterioration. On the other hand, cellulosic biomass can often be bulky and would require a welldeveloped transportation infrastructure for delivery to processing plants after harvest.

Significant technological challenges still need to be overcome to make the production of ethanol from lignocellulosic feedstocks commercially competitive. It is still uncertain when conversion of cellulosic biomass into advanced fuels may be able to contribute a significant proportion of the world's liquid fuels. Currently, there are a number of pilot and demonstration plants either operating or under development around the world. The speed of expansion of biochemical and thermochemical conversion pathways will depend upon the development and success of pilot projects currently under way and sustained research funding, as well as world oil prices and private-sector investment.

In summary, second-generation biofuels based on lignocellulosic feedstocks present a completely different picture in terms of their implications for agriculture and food security. A much wider variety of feedstocks could be used, beyond the agricultural crops currently used for firstgeneration technologies, and with higher energy yields per hectare. Their effects on commodity markets, land-use change and the environment will also differ – as will their influence over future production and transformation technologies.

Potential for Bioenergy

The technical and economic potential for bioenergy should be discussed in the context of the increasing shocks and stress on the global agriculture sector and the growing demand for food and agricultural products that is a consequence of continuing population and income growth worldwide. What is technically feasible to produce may not be economically feasible or environmentally sustainable.

Because bioenergy is derived from biomass, global bioenergy potential is ultimately limited by the total amount of energy produced by global photosynthesis. Plants collect a total energy equivalent of about 75 000 Mtoe (3 150 Exajoule) per year – or six to seven times the current global energy demand. However, this includes vast amounts of biomass that cannot be harvested. In purely physical terms, biomass represents a relatively poor way of harvesting solar energy, particularly when compared with increasingly efficient solar panels.

A number of studies have gauged the volume of biomass that can technically contribute to global energy supplies. Their estimates differ widely owing to different scopes, assumptions and methodologies, underscoring the high degree of uncertainty surrounding the possible contribution of bioenergy to future global energy supply. The last major study of bioenergy conducted by the International Energy Agency (IEA) assessed, on the basis of existing studies, the range of potential bioenergy supply in 2,050 from a low of 1,000 Mtoe to an extreme of 26,200 Mtoe. The latter figure was based on an assumption of very rapid technological progress; however, the IEA indicates that a more realistic assessment based on slower yield improvements would be 6,000–12,000 Mtoe. A mid-range estimate of around 9,500 Mtoe would, according to the IEA, require about onefifth of the world's agricultural land to be dedicated to biomass production.

More important than the purely technical viability is the question of how much of the technically available bioenergy potential would be economically viable. The long-term economic potential depends crucially on assumptions concerning the prices of fossil energy, the development of agricultural feedstocks and future technological innovations in harvesting, converting and using biofuels.

A different way of looking at the potential for biofuel production is to consider the relative land-use requirements. In its "Reference Scenario" for 2030 in World Energy Outlook 2006, the IEA projects an increase in the share of the world's arable land devoted to growing biomass for liquid biofuels from 1 percent in 2004 to 2.5 percent in 2030. Under its "Alternative Policy Scenario", the share in 2030 increases to 3.8 percent. In both cases, the projections are based on the assumption that liquid biofuels will be produced using conventional crops. Should second-generation liquid biofuels become widely commercialized before 2030, the IEA projects the global share of biofuels in transport demand to increase to 10 percent rather than 3 percent in its Reference Scenario and 5 percent in the Alternative Policy Scenario. Land-use requirements would go up only slightly, to 4.2 percent of arable land, because of higher energy yields per hectare and the use of waste biomass for fuel production. Nevertheless, this illustrates that, even under a second-generation scenario, a hypothetical large-scale substitution of liquid biofuels for fossil-fuel-based petrol would require major conversion of land.

The potential for current biofuel technologies to replace fossil fuels is also illustrated by a hypothetical calculation by Rajagopal et al. They report theoretical estimates for global ethanol production from the main cereal and sugar crops based on global average yields and commonly reported conversion efficiencies. The results of their estimates are summarized in table below. The crops shown account for 42 percent of total cropland today. Conversion of the entire crop production to ethanol would correspond to 57 percent of total petrol consumption. Under a more realistic assumption of 25 percent of each of these crops being diverted to ethanol production, only 14 percent of petrol consumption could be replaced by ethanol. The various hypothetical calculations underline that, in view of their significant land requirements, biofuels can only be expected to lead to a very limited displacement of fossil fuels. Nevertheless, even a very modest contribution of biofuels to overall energy supply may yet have a strong impact on agriculture and on agricultural markets.

Hypothetical potential for ethanol from principal cereal and sugar crops

CROP	GLOBAL AREA	GLOBAL PRODUCTION	BIOFUEL YIELD	MAXIMUM ETHANOL	PETROL EQUIVALENT	SUPPLY AS SHARE OF 2003 GLOBAL PETROL USE[1]
	(Million ha)	(Million tonnes)	(Litres/ha)	(Billion litres)	(Billion litres)	(Percentage)
Wheat	215	602	952	205	137	12
Rice	150	630	1 806	271	182	10
Maize	145	711	1 960	284	190	17
Sorghum	45	59	494	22	15	1
Sugar cane	20	1 300	4 550	91	61	6
Cassava	19	219	2 070	39	26	2
Sugar beet	5.4	248	5 060	27	18	2
Total	**599**	...	...	**940**	**630**	**57**

Wind Turbines

Farmers have been using traditional windmills for generations to pump water and mill grain. Now, they are hosting modern wind turbines on their farmland to generate electricity.

Wind Turbines are a Stable Source of Income

The money they earn from leasing their land to wind companies is a stable source of income. This can help protect the farm from fluctuations in commodity prices or poor crop yields during drought years. It can also allow them to reinvest in their farming business by purchasing new equipment or making other improvements.

Turbines do not Restrict Farming Operations

Farmers are able to continue farming the majority of their land:

- After the turbines and related infrastructure are installed, farming can continue nearly right to the base of the turbine.
- Each turbine only removes about ¼ to ½ of an acre from agricultural production.
- A typical wind farm leaves 98 percent of land undisturbed, meaning it's free for other uses like farming and ranching.
- Often times the construction process will including creating new roads or updating existing roads, which can improve operations.
- If crops are damaged during construction, wind developers will typically reimburse the landowner for the lost revenue.

Drain Tiles and Drainage

The construction of a wind farm typically leaves a minimal footprint. Developers plan ahead to minimize damage to drain tiles by mapping the network ahead of time and avoiding tiles whenever possible. However, the operation of bulldozers and cranes during construction can occasionally damage drain tiles.

Wind developers partner with farmers and other landowners through the process of constructing and operating a wind farm to ensure that any drain tile damage is repaired, and provision is often usually covered in a landowner's lease agreement with the developer.

Crop Dusting

Every day crop dusting pilots routinely fly near trees, power poles and wires, and other obstacles. Wind turbines are another structure to navigate around and may require some additional maneuvering to accommodate, but do not represent significant new challenges when compared to these other typical obstacles. Regardless, some crop dusting companies may charge extra for servicing land with turbines, or will designate specific pilots who have experience for jobs on wind farms.

Landowners have the ability to negotiate their lease with a wind farm developer to include specific provisions related to crop dusting, and wind developers typically employ various measures to ensure that landowners can continue to manage their farming operations effectively.

Automatic Nursery Seeding Machine

Seeding machine is mainly used for planting seedlings such as rice, tomatoes, watermelon, cucumber, pumpkin, onion, rapeseed, and pepper. Rice nursery sowing machine can automatically fill the soil into the nursery tray and accurately drops the seed into the middle of the tray.

The seedling can grow quickly without any damage to roots. Most importantly, the machine has a high yield and is suitable for long-distance transportation:

- Soil container, Nursery tray, Soil laying motor, Brush adjustment, Speed adjustment, Falling soil adjustment, Belt, Brush height adjustment, Conveyor motor and Soil brushing device.
- Brush motor, Flat brush, Conveyor belt, Trundle, Air inflow switch, Precision pressure regulating filter, Air pressure regulating valve, Barometer and Conveyor belt adjustment.
- Conveyor motor, Earth pressing device, Cylinder, Cylinder adjustment, Air blowing device, Seed container adjustment, Seeds suction pressure regulation, Seed dropping pressure regulation, Vibration adjustment, Pressure gauge, Wheel and Height adjustment
- Belt adjustment screw, Belt adjustment screw, Brush adjustment, Speed adjustment, Falling soil adjustment, Conveyor motor, Nursery tray, Falling soil motor, Brush, Brush motor and Soil container.

The original meaning of cultivating seedlings is to do it in nurseries, hotbeds or greenhouses so as to transplant them into the land for planting. It can also refer to the stage where various organisms are artificially protected until they can survive independently.

Through observation, it is found that the existing nursery seeder machine generally has such problems as a single structure, low seedling efficiency, and complex operation, which brings certain limitations in the actual seedling raising process.

Therefore, how to provide a well-structured seeding nursery machine with high efficiency and simple and quick operation has become an important issue, and it needs to be solved at present.

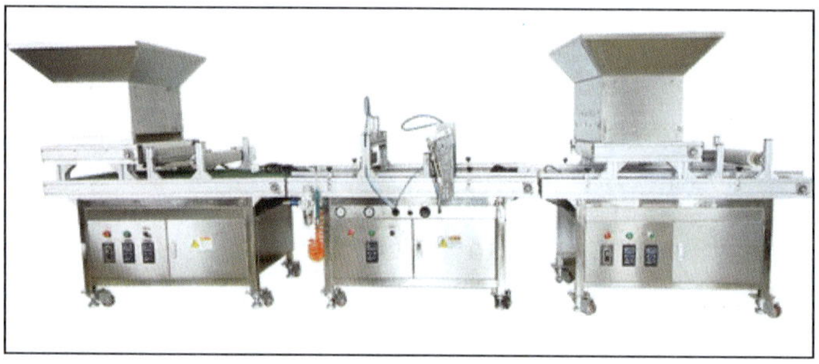

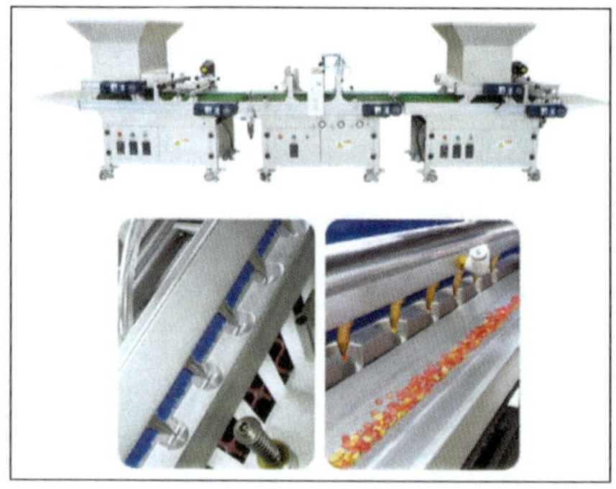

The working principle of the automatic nursery seeding machine:

1. Firstly, the operator places the seedling tray into the conveyor belt. The photoelectric sensor covers the soil aftcr sensing the position of the tray.
2. Then brush sweeps the soil of the seedling tray.
3. The seedling tray passes through hole digging area that presses into a single hole.
4. When the seedling tray reaches the seeding machine, it uses a row of nozzles to adsorb the seed from the vibrating plate, and the nozzle releases the seed into the middle of the tray.
5. Seeds fall on a tray through the dropping tube, and then the nozzle automatically repeats the same action.
6. Finally, the seedling tray is brushed again by soil.

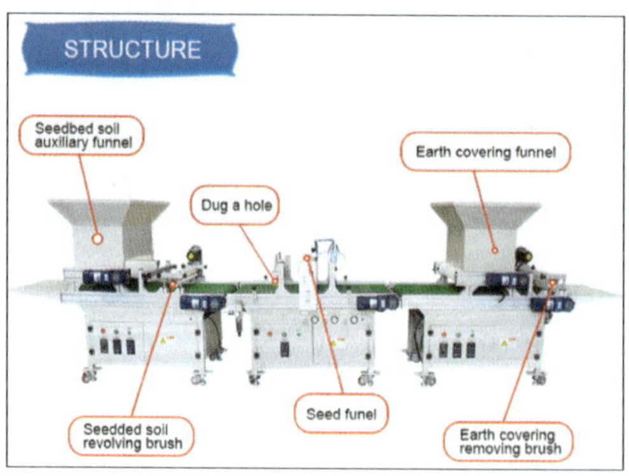

Automatic Nursery Seeding Machine Advantage

1. The single-particle of the needle-type nozzle is high and can be applied to many fields.
2. Users can configure the corresponding seeding board and a suction nozzle according to different seeds and sowing requirements.

3. Nursery tray seeder is fit for such seeds as follow:
    - The larger size seeds: pumpkin, peas, corn, watermelon, etc.,
    - The smaller size seeds: petunia, celery, cabbage, etc.,
    - Normal size seeds: pepper, tomato, etc.
4. There is a hole at the middle of tray, which is good for ventilation.
5. The size of tray is various such as 4*8, 5*10, 6*12 , we can choose them according to your needs.
6. The seedling bears high survival rate after cultivating by this sowing machine.
7. It is easy to change the needle-type nozzle to fit for seeds with different size.
8. Seeding machine is made of stainless steel, which is corrosion resistance and boasts long service life.

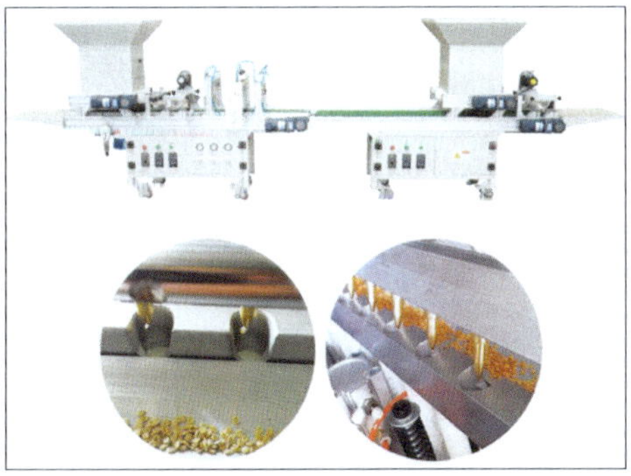

Nursery Seeding Machine

1. Seeding accuracy: 97% or more (pelleted seeds), imported electrical accessories.
2. With automatic photoelectric sensing detection system, the seeding machine won't load soil and drop seeds if there is no tray.

3. A dedicated seed recovery device that keeps seeds from being wasted.
4. Nursery seeding machine is suitable for seeds between 0.3-12mm, and the shape of the seed is not limited, so the scope of application is wide.
5. With a photoelectric sensing system, the soil covering process can be conducted automatically.
6. Simply adjust the upper and lower partitions to meet the required amount of soil for the trays.
7. Applicable maximum tray width (mm): 540.
8. Conveyor device is equipped with a frequency conversion adjustment system.
9. Nursery sowing machine uses button operating system, and soil covering, soil pressing, and seeding are simple, practical and fast.
10. The soil loading and conveying device is equipped with a variable frequency speed control system, which perfectly matches the speed of the next process.
11. Seeding speed: 400-550 trays per hour.

References

- What-is-precision-agriculture: sustainableamerica.org, Retrieved 2 June, 2019
- Precision-agriculture-precision-farming: techtarget.com, Retrieved 22 May, 2019
- Controlled-Traffic-Farming: researchgate.net, Retrieved 23 February, 2019
- Autonomous-farm-equipment-equals-higher-production: agritechtomorrow.com, Retrieved 10 July, 2019
- Autonomous, newsletters: agriculture.purdue.edu, Retrieved 13 March, 2019
- Autonomous-tractors: precisionagricultu.re, Retrieved 29 June, 2019
- Calibration-of-grain-seed-drills: extension.psu.edu, Retrieved 18 January, 2019
- Seed-processing-equipments., seed-certification: agritech.tnau.ac.in, Retrieved 30 January, 2019
- Disc-harrows-powerful-farm-equipment-for-soil-preparation: agrivi.com, Retrieved 31 March, 2019
- Agriculture-and-biofuels, agriculture: epa.gov, Retrieved 14 July, 2019
- Agriculture, wind-in-my-community, benefits-of-wind, wind: awea.org, Retrieved 17 May, 2019
- Farm-seeding-machine: agriculture-machine.com, Retrieved 19 April, 2019

3

Remote Sensing in Agriculture

Remote sensing is a process that involves the acquisition of information about an object or phenomenon without making physical contact with the object. It is used in horticulture, assessment of crop damage, crop identification, etc. This chapter has been carefully written to provide an easy understanding of the applications of remote sensing in agriculture.

Remote Sensing Applications in Agriculture

Remote sensing is the acquisition of information about an object or any phenomenon without making any physical contact with the object. It is a phenomenon that has numerous applications including photography, surveying, geology, forestry and many more. But it is in the field of agriculture that remote sensing has found significant use.

There are very many applications of remote sensing in the agricultural sector. Below is a summary of these applications:

- Crop production forecasting: Remote sensing is used to forecast the expected crop production and yield over a given area and determine how much of the crop will be harvested under specific conditions. Researchers can be able to predict the quantity of crop that will be produced in a given farmland over a given period of time.
- Assessment of crop damage and crop progress: In the event of crop damage or crop progress, remote sensing technology can be used to penetrate the farmland and determine exactly how much of a given crop has been damaged and the progress of the remaining crop in the farm.
- Horticulture, Cropping Systems Analysis: Remote sensing technology has also been instrumental in the analysis of various crop planting systems. This technology has mainly been in use in the horticulture industry where flower growth patterns can be analyzed and a prediction made out of the analysis.
- Crop Identification: Remote sensing has also played an important role in crop identification especially in cases where the crop under observation is mysterious or shows some mysterious characteristics. The data from the crop is collected and taken to the labs where various aspects of the crop including the crop culture are studied.
- Crop acreage estimation: Remote sensing has also played a very important role in the estimation of the farmland on which a crop has been planted. This is usually a cumbersome procedure if it is carried out manually because of the vast sizes of the lands being estimated.
- Crop condition assessment and stress detection: Remote sensing technology plays an important role in the assessment of the health condition of each crop and the extent to

which the crop has withstood stress. This data is then used to determine the quality of the crop.

- Identification of planting and harvesting dates: Because of the predictive nature of the remote sensing technology, farmers can now use remote sensing to observe a variety of factors including the weather patterns and the soil types to predict the planting and harvesting seasons of each crop.
- Crop yield modelling and estimation: Remote sensing also allows farmers and experts to predict the expected crop yield from a given farmland by estimating the quality of the crop and the extent of the farmland. This is then used to determine the overall expected yield of the crop.
- Identification of pests and disease infestation: Remote sensing technology also plays a significant role in the identification of pests in farmland and gives data on the right pests control mechanism to be used to get rid of the pests and diseases on the farm.
- Soil moisture estimation: Soil moisture can be difficult to measure without the help of remote sensing technology. Remote sensing gives the soil moisture data and helps in determining the quantity of moisture in the soil and hence the type of crop that can be grown in the soil.
- Irrigation monitoring and management: Remote sensing gives information on the moisture quantity of soils. This information is used to determine whether a particular soil is moisture deficient or not and helps in planning the irrigation needs of the soil.
- Soil mapping: Soil mapping is one of the most common yet most important uses of remote sensing. Through soil mapping, farmers are able to tell what soils are ideal for which crops and what soil require irrigation and which ones do not.
- Monitoring of droughts: Remote sensing technology is used to monitor the weather patterns including the drought patterns over a given area. The information can be used to predict the rainfall patterns of an area and also tell the time difference between the current rainfall and the next rainfall which helps to keep track of the drought.
- Land cover and land degradation mapping: Remote sensing has been used by experts to map out the land cover of a given area. Experts can now tell what areas of the land have been degraded and which areas are still intact. This also helps them in implementing measures to curb land degradation.
- Identification of problematic soils: Remote sensing has also played a very important role in the identification of problematic soils that have a problem in sustaining optimum crop yield throughout a planting season.
- Crop nutrient deficiency detection: Remote sensing technology has also helped farmers and other agricultural experts to determine the extent of crop nutrients deficiency and come up with remedies that would increase the nutrients level in crops hence increasing the overall crop yield.
- Reflectance modeling: Remote sensing technology is just about the only technology that can provide data on crop reflectance. Crop reflectance will depend on the amount of moisture

in the soil and the nutrients in the crop which may also have a significant impact on the overall crop yield.

- Determination of water content of field crops: Apart from determining the soil moisture content, remote sensing also plays an important role in the estimation of the water content in the field crops.
- Crop yield forecasting: Remote sensing technology can give accurate estimates of the expected crop yield in a planting season using various crop information such as the crop quality, the moisture level in the soil and in the crop and the crop cover of the land. When all of this data is combined it gives almost accurate estimates of the crop yield.
- Flood mapping and monitoring: Using remote sensing technology, farmers and agricultural experts can be able to map out the areas that are likely to be hit by floods and the areas that lack proper drainage. This data can then be used to avert any flood disaster in future.
- Collection of past and current weather data: Remote sensing technology is ideal for collection and storing of past and current weather data which can be used for future decision making and prediction.
- Crop intensification: Remote sensing can be used for crop intensification that includes collection of important crop data such as the cropping pattern, crop rotation needs and crop diversity over a given soil.
- Water resources mapping: Remote sensing is instrumental in the mapping of water resources that can be used for agriculture over a given farmland. Through remote sensing, farmers can tell what water resources are available for use over a given land and whether the resources are adequate.
- Precision farming: Remote sensing has played a very vital role in precision agriculture. Precision agriculture has resulted in the cultivation of healthy crops that guarantees farmers optimum harvests over a given period of time.
- Climate change monitoring: Remote sensing technology is important in monitoring of climate change and keeping track of the climatic conditions which play an important role in the determination of what crops can be grown where.
- Compliance monitoring: For the agricultural experts and other farmers, remote sensing is important in keeping track of the farming practices by all farmers and ensuring compliance by all farmers. This helps in ensuring that all farmers follow the correct procedures when planting and when harvesting crops.
- Soil management practices: Remote sensing technology is important in the determination of soil management practices based on the data collected from the farms.
- Air moisture estimation: Remote sensing technology is used in the estimation of air moisture which determines the humidity of the area. The level of humidity determines the type of crops to be grown within the area.
- Crop health analysis: Remote sensing technology plays an important role in the analysis of crop health which determines the overall crop yield.

- Land mapping: Remote sensing helps in mapping land for use for various purposes such as crop growing and landscaping. The mapping technology used helps in precision agriculture where specific land soils are used for specific purposes.

Agriculture Sensors

Sensors used in smart farming are known as agriculture sensors. These sensors provide data which assist farmers to monitor and optimize crops by adapting to changes in the environmental conditions. These sensors are installed on weather stations, drones and robots used in the agriculture industry. They can be controlled using mobile apps specifically developed for the purpose. Based on wireless connectivity either they can be controlled directly using wifi or through cellular towers with cellular frequencies with the help of mobile phone app.

The table below mentions list of agriculture Sensors used for different functions in agricultural industry.

Agriculture Sensors	Functional Description
Location Sensors	These sensors determine latitude, longitude and altitude of any position within required area. They take help of GPS satellites for this purpose.
Electro-Chemical Sensors	These sensors help in gathering chemical data of the soils by detecting specific ions in the soil. They provide information in the form of pH and soil nutrient levels.
Mechanical Sensors	These sensors are used to measure soil compaction or mechanical resistance.
Air Flow Sensors	These sensors are used to measure air permeability. They are used in fixed position or in mobile mode.

Uses of Agriculture Sensors

Following are the uses of Agriculture Sensors:

- They are used in agricultural weather stations: These equipments are equipped with sensors which provide informations such as soil temperature at various depths, air temperature, rainfall, leaf wetness, chlorophyll, wind direction, solar radiation, relative humidity, atmospheric pressure etc.

- They are used in many equipments (e.g. dendrometer) developed by agro based industries for agricultural or farming applications such as measuring trunk diameter, leaf wetness and so on.
- They are used in agriculture drones for the purpose of spraying insecticides and pesticides.
- Solar based pumps which are mobile operated have become very popular due to reduction in cost to electricity.
- E-fences have become popular in rural INDIA which helps save crops from animals such as elephants.

Benefits or advantages of Agriculture Sensors

Following are the benefits or advantages of Agriculture Sensors:

- They are invented to meet increasing demand of food by maximizing yields with minimum resources such as water, fertilizers and seeds. They fulfill this by conserving resources and mapping fields.
- They are simple to use and easy to install.
- They are cheaper.
- In addition to agricultural use, they can also be used for pollution and global warming.
- They are equipped with wireless chip so that they can be remotely controlled.

Drawbacks or Disadvantages of Agriculture Sensors

Following are the drawbacks or disadvantages of Agriculture Sensors:

- Smart farming and IoT technology require continuous internet connectivity. This is not available in developing coutries such as INDIA and other part of the world.
- There is presumption in the market that consumers are not always ready to adopt latest IoT devices equipped with agriculture sensors.
- The basic infrastructure requirements such as smart grids, traffic systems and cellular towers are not available everywhere. This further hinders the growth of its use.

Key Issues

Gradually it has been realized by both agriculture scientists and large crop farm owners that sensor systems are the ideal allied tools to rely on? in real time at the right stage. Smart farming through sensorization is the best solution with a number of issues to focus on, such as:

- Determination of potential applications,
- Exposure to continuous moisture conditions,
- Assessing requirement of wired or wired sensors,

- In case of wireless sensors it is important to determine the range/proximity,
- Determining the source of power,
- Data streaming (gathering) frequency,
- Analytical need,
- Sensor interfacing mode,
- IoT Connectivity and remote monitoring need.

Main Needs

From the point of view of an agriculture farm owner the main requirements are:

- Monitoring crop yield,
- Soil analysis,
- Monitoring soil moisture,
- Estimating crop water requirements,
- Disease detection,
- Detecting fruit/crop ripening,
- Crop readiness,
- Detection of fruit ripening,
- Auto-steer and GPS guidance,
- Monitoring and control of sprayers,
- Row control on planters and seeders,
- CO_2 and Oxygen monitoring and control of enclosed green houses,
- Fertilizer analysis,
- Weed monitoring and control,
- Chlorophyll estimation,
- Disease Detection and Diagnosis,
- Water need estimation,
- Yield estimation and prediction.

Smart Approach

The path toward complete integration is likely to be achieved through:

- Sensor fusion,
- Combinational sensors,

- Developing newer sensor types,
- Robotics,
- Remote Sensing,
- In-field Sensing,
- Data Management,
- Variable Rate Applications,
- Telematics,
- In-field Sensing,
- Artificial intelligence (AI).

Key Sensors, Robotics and Artificial Intelligence

A list of sensors covering the above topics, but are not limited to, is the following:

- Biological (including chemical and gas analyzers),
- Water sensors,
- Meteorological sensors,
- Weed seekers,
- Optical sensors – Hyperspectral, Multispectral, Fluorescence and thermal sensing,
- Optical cameras,
- Light Detection and Ranging (LIDAR),
- Sensors for detection of Microorganisms and Pest management,
- Photometric sensors,
- Soil respiration or moisture,
- Photosynthesis sensors,
- Leaf Area index (LAI) sensors,
- Accelerometers – Helps determine Leaf Angle Index. Also used as an equipment rollover alarm,
- Range finders,
- Dendrometers,
- Hygrometers,
- Temperature sensors,
- Gas sensors,

- Chlorophyll meters,
- Image sensors – for disease detection,
- Volatile components detection sensors, and electronic noses and tongues,
- Sensors for positioning, navigation, and obstacle detection,
- Level sensors – for crop storages silos,
- LiDAR, Radar, Laser, Sonar – for accident avoidance,
- Position sensors – sprayers, harvesters, sowing,
- GPS (for geolocation-GNSS receivers, RTK signals) – can accurately position a moving vehicle within 30cm or less using GPS,
- Air flow sensors – (measure measure soil air permeability).

Modern agriculture involves usage of large number of agricultural implements from sowing to crop harvesting and storages. Using a large number of sensors makes the entire operation smart and cost effective due to high precision. The agriculture sector has been experiencing automation and robotic equipment are playing crucial operational precision. Autonomous tractors, robotic harvesters, robotic weeders and robotic data scouts currently complement farm equipment. Aerial imagery is another operational technique where large farm areas are monitored. Satellites, planes and recently unmanned drones are deployed for areal imagery operations

Soil and Crop Sensing

Sensors that measure a variety of essential soil properties on the go are being developed. These sensors can be used either to control variable rate application equipment in real-time or in conjunction with a Global Positioning System (GPS) to generate field maps of particular soil properties. Depending on the spacing between passes, travel speed, and sampling and/or measurement frequency, the number of measurement points per acre varies; however, in most cases, it is much greater than the density of manual grid sampling. The cost of mapping usually is reduced as well.

Measuring Soil Properties

When thinking about an ideal precision agriculture system, producers visualize a sensor located in direct contact with, or close to, the ground and connected to a "black box" which analyzes sensor response, processes the data, and changes the application rate instantaneously. They also hope that the real-time information detected by the sensor and used to prescribe the application rate would optimize the overall economic or agronomic effect of the production input. This approach, however, does not take into account several difficulties met in the "real world":

- Most sensors and applicator controllers need a certain time for measurement, integration, and/or adjustment, which decreases the allowable operation speed or measurement density.
- Variable rate fertilizer and pesticide applicators may need additional information (like yield potential) to develop prescription algorithms (sets of equations).
- Currently, there is no site-specific management prescription algorithm proven to be the most favorable for all variables involved in crop production.

Rather than using real-time, on-the-go sensors with controllers, a map-based approach may be more desirable because of the ability to collect and analyze data, make the prescription, and conduct the variable rate application in two or more steps. In this case, multiple layers of information including yield maps, a digital elevation model (DEM), and various types of imagery could be pooled together using a geographic information system (GIS) software package designed to manage and process spatial data. Prescription maps can be developed using algorithms that involve several data sources as well as personal experience.

Sensor Data usage

Although various vehicle-based soil sensors are under development, only electromagnetic sensors are commercially available and widely used. Ideally, producers would like to operate sensors that provide inputs for existing prescription algorithms. Instead, commercially available sensors provide measurements such as electrical conductivity (EC) that cannot be used directly since the absolute value depends on a number of physical and chemical soil properties such as: texture, organic matter, salinity, moisture content, etc. Alternatively, electromagnetic sensors give valuable information about soil differences and similarities, which makes it possible to divide the field into smaller and relatively consistent areas referred to as management zones.

For example, such zones could be defined according to various soil types in a field. In fact, electrical conductivity maps usually can better reveal boundaries of certain soil types than soil survey maps (used for rural property tax assessment). Different anomalies such as eroded hillsides or ponding also can be easily identified on an electrical conductivity map. The figure below compares a soil survey and an electrical conductivity map for the same field showing some differences in boundaries.

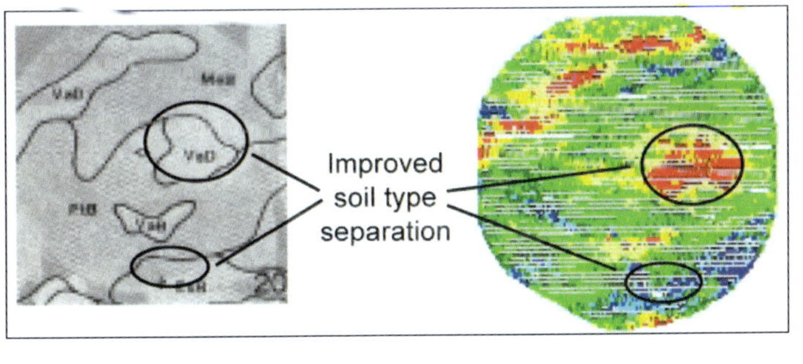

Yield maps also frequently correlate to electrical conductivity maps, as shown below. In many instances, such similarities can be explained through differences in soil. In general, the electrical conductivity maps may indicate areas where further exploration is needed to explain yield differences. Both yield potential and nutrient availability maps may have a similar pattern as soil texture and/or organic matter content maps. Often these patterns also can be revealed through an electrical conductivity map.

Therefore, it seems reasonable to use on-the-go mapping of electromagnetic soil properties as one layer of data to discover the heterogeneity (differences) of soil within a field (similar to using bare soil imagery). Zones with similar electrical conductivity and a relatively stable yield may receive a uniform treatment that can be prescribed based on fewer soil samples in the zones on the electrical conductivity map.

As new on-the-go soil sensors are developed, different real-time and map-based variable rate soil treatments may be economically applied to much smaller field areas, reducing the effect of soil variability within each management zone.

Optical Sensors

Optical sensors use light reflectance to characterize soil. These sensors can simulate the human eye when looking at soil as well as measure near-infrared, mid-infrared, or polarized light reflectance. Vehicle-based optical sensors use the same principle technique as remote sensing.

There is a very wide range of optical sensors applied in agriculture, which goes from sensors used to analyze soil attributes to sensors installed in combines to measure protein content in wheat grains while they are being harvested.

This kind of optical sensors began to be studied in 1991 with the development of sensors focused in weed detection at the Oklahoma State University. Just based on the simple fact that soil and plants (weeds) have a different interaction with the light emitted from the sensors, allowing identifying what is soil and what is a plant.

In 1992 there was the first discussion between the Departments of Plant and Soil Sciences and Biosystems and Agricultural Engineering concerning the possibility of sensing biomass in wheat and bermudagrass. The objective was to use biomass as an indicator of nutrient need (based on removal). In 1993 Dr. John Solie, Dr. Marvin Stone and Shannon Osbourne collected sensor readings at ongoing bermudagrass, with nitrogen rates versus nitrogen timing experiments with the Noble Foundation in Ardmore, Oklahoma. Initial results were promising enough to continue this work in wheat. And in fall of 1993 the first variable application of nitrogen was done across a 70 meter transect. In 1994 John Ringer and Shannon Osbourne collected sensor readings and later applied variable N fertilizer rates based on the first bermudagrass algorithm developed by TEAM-VRT.

In the subsequence years the research advanced creating algorithms for nitrogen application in many different crops. And nowadays we have commercial sensors being sold to farmers to make real-time application of nitrogen, growth-regulators and desiccants.

The Types of Sensors used

The sensors to measure crops reflectance can be classified according to the platform, like satellites, aerial (airplanes, UAV's - unmanned aerial vehicles, balloon) and ground based. For satellites, airplanes and UAV's it is most common to use cameras to collect images, and ground based optical sensors can collect reflectance data and storage in a text file. The ground sensors can also be classified into active or passive. The basic difference is that the passive sensors need an external source of light, like the sun. The active sensors have its own source of light, which can be a wide range light or a specific wavelength.

There are available a few brands on the market, each one with its own construction characteristics like internal batteries, GPS antenna, data logger and log frequency. The table has some examples:

Table: Optical sensors and manufacturers.

Manufacturer	Sensor	Country
AgLeader	OptRx	United States
Falker	ClorofiLOG	Brazil
Force A	Dualex and Multiplex	France
Fritzmeier	ISARIA	Germany
Holland Scientific	CropCircle	United States
Konica Minolta	SPAD	Japan
Topcon	CropSpec	Japan
Trimble	GreenSeeker	United States
Yara	N-Sensor	Germany

There are also some differences about the distance from the sensor to the crops. For example, the SPAD and the ClorofiLOG sensors need to make static measurements, touching the sensors on crop's leaves. The other sensors make the measurements from centimeters to meters, but do not need to have contact with the leaves.

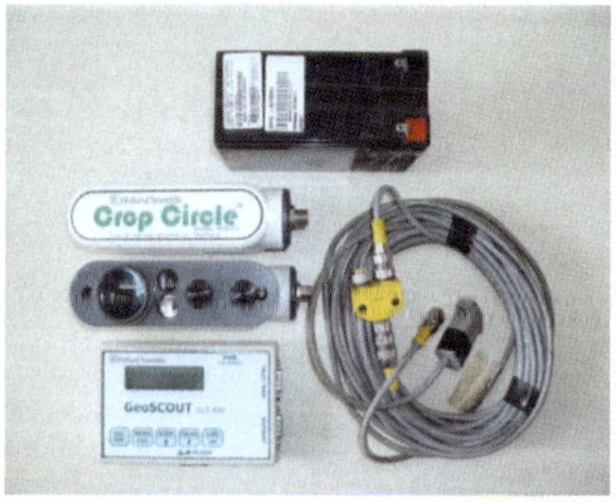

Crop Circle optical sensor

The Crop Circle ACS-210 sensor, manufactured by the Holland Scientific Inc., Lincoln, Nebraska. As shown in the picture, this sensor has one LED (Light Emitting Diode) that emits active radiation simultaneously in visible and near infrared light with a system called PolySource and two silicon photodiodes with a spectral range of 320 to 1,100 nm to detect light. One detector works between 400 and 680 nm and the other between 800 and 1,100 nm. Using a filter on each detector, the wavelengths used are the amber light (590 ± 5 nm) and near infrared (880 ± 10 nm). The sensor must be placed between 0.25 and 2.13 m from the target, and then the light reaches the target and reflects part of the energy, which is received by the detectors. This sensor needs external batteries (12 V) and GPS antenna, but has its own data logger, that saves the data in a SD card. Also in figure below there is an example on how to install the sensors in a motorcycle to collect data on-the-go.

The figure below has two other examples, one is the Hand held GreenSeeker, from Trimble, and the other one is a prototype from Oklahoma State University, which now is also being commercialized by Trimble. The Pocket Sensor was designed for small farmers, made to be a low cost sensor, it does not use GPS signal or a data logger, and you can just position the sensor over the crop, press the trigger and see the value of NDVI (Normalized Difference Vegetation Index).

GreenSeeker sensor collecting NDVI values for weed (a) and (b); detailed pictures from GreenSeeker Hand Held (c) and Pocket Sensor (d); and GreenSeeker mounted on a sprayer (e) and on a motorcycle (f) collecting NDVI values in sugarcane.

The Range in the Spectrum used by Sensors and Response on Crops. The electromagnetic spectrum goes from gamma rays to radio waves. These sensors used for measure crop reflectance usually work in the visible and near infrared region of the spectrum, and combining at least two wavelengths to calculate vegetation indices. From the agronomic point of view, the visible light has a straight relationship with the chlorophyll content, absorbing the blue and red lights, and reflecting the green light. That's what makes us to see healthy plants as green. The near infrared light, not visible by the human eye, is reflected by the mesophyll cells, which is found in more quantity in a plant than chlorophyll, resulting in a much higher reflectance than visible lights. Using both wavelengths it is possible to evaluate the color and biomass production of a crop. In practice, greener and higher biomass plants have a higher chance to have higher yields.

Remote sensing can be defined as the technique of acquisition and application of information about an object without any physical contact. The information is acquired by the detection and measurement of changes that an object imposes to the environment around it, and this signal may include an electromagnetic field emitted and/or reflected, acoustic waves reflected and/or disturbed by the object or the disturbances of gravitational field or magnetic potential with the presence of the object. Usually the acquisition of information is based on capturing the electromagnetic signals that cover the entire spectrum of electromagnetic waves to radio long waves, passing through microwaves, sub millimeter, thermal and near infrared, visible, ultraviolet, X ray and gamma ray.

The sensors used by remote sensing are devices capable of detect and register electromagnetic radiation in certain range of electromagnetic spectrum and generate information that may be transformed in a product, which can be interpreted, like an image, graphic or tables. The sensor systems are basically formed by an optical part, constructed with lens or mirrors that have the objective of receiving and directing the energy from the targets to the detectors.

The spectral reflectance measurements are the non-destructive approach most promising for determining nitrogen deficiency in crops. For this purpose, remote sensing has been used to evaluate crop conditions related to nitrogen. Many researchers used remote sensing to estimate crop parameters like LAI (Leaf Area Index), leaf chlorophyll content, soil cover, dry matter, water content, yield, nitrogen content and many others.

In reference studying spectral signature of green leaves, was found that wavelengths from 400 to 700 nm (visible), the reflectance is low, about 10%, with a smooth increase at 550 nm (green). In the near infrared region (700 to 1,300 nm) there is another increase in the reflectance, close to 50%. For the visible light the low reflectance is related to the absorption associated to leaf pigments, mainly chlorophyll. And the reflectance increase at the near infrared region is due the internal structure of the leaves (size and shape of the cells and empty spaces). Combining both visible and near infrared reflectance there are the vegetation indices, which can be resulted from two or more spectral bands.

First, it was proposed the ratio between the measurements from 800 and 675 nm to determine the leaf area index in forests. The relation between these two wavelengths is known as RVI (Ratio Vegetation Index). The NDVI (Normalized Difference Vegetation Index) appeared just after, which was found a relation between two wavelengths that better solved the issues about soil interference on the vegetation measurements, and also reduced the atmosphere influence and sun angle variations.

The normalization proposed guarantees that the values obtained from the NDVI are contained in the same scale of values, between -1 and 1, as shown in the equation:

$$NDVI = \frac{(\rho_{IR} - \rho_V)}{(\rho_{IR} + \rho_V)}$$

Which:

ρ_{IR} = infrared reflectance;

ρ_V = visible reflectance.

The normalization is produced by the combination of the strong absorption by the chlorophyll in the red region and the strong reflectance in the near infrared, due to the dispersion in the leaf mesophyll and the absence of absorption by the pigments. The peculiarity assigned to the NDVI is the early saturation, what makes it insensible to biomass increase after certain development stage. That means the NDVI stabilizes, showing constant values, even with the biomass increase.

Aspects of use in the Field and what can affect the Measurements

Measurements in the field can be affected by the sensor positioning, like the distance from the crop, dependence of a light source, the presence of dew over the leaves and also because of factors that can stress the plants. The light source classifies the sensors in active and passive sensors. The passive sensors are dependent of sunlight, not working at night or might show different readings when there are clouds or shadows. There is a limitation of the distance from the target to the sensor, because if it is too close the sensor may not capture the reflectance, and if it is too far the data may have noise signals. And the presence of dew is just because the presence of water over the leaves can change the reflectance in both visible and near infrared. With the presence of dew the reflectance increases, but as visible light is more affected, consequently reduces the NDVI values.

Other methods can be used instead of on-the-go sensing to indirectly measure nitrogen stress, like chlorophyll content in the leaves using chlorophyll meters like SPAD and ClorofiLOG. In reference, analyzing aerial and satellite images it was found lower correlation with wheat crops variables then ground sensors, besides depending on the weather it is not possible to have satellite images due the presence of clouds.

The spectral data that can be obtained from satellite images have low temporal resolution to use in agriculture. The ground acquisition is independent from the weather, the data collection can be done together with other machinery operation and the data are available just after you finish, with no need of complicated processing. Higher correlation was found between the NDVI obtained from the GreenSeeker then the NDVI obtained from the satellite Quickbird II images for the amount of nitrogen applied, nitrogen content on the flag leaf, yield and protein content in wheat grains.

In reference the author used a sensor installed in the front of a tractor to acquire spectral data in wheat. The tractor was at the speed of 0.8 m/s, and the sensor was set to register data with the frequency of 10 measurements per second. The authors found good relationship between the reflectance and nitrogen absorption by the wheat crop. The same configuration was used to collect data in pasture and also found good relationship between the NDVI and the nitrogen removal by the crop. But depending on the sensor used, the speed can also affect the NDVI, increasing the coefficient of variation of the measurements.

An experiment conducted along the day, collecting NDVI values at the same spot but in different times of the day showed that the presence of dew on the leaves reduced the NDVI values of 12% for GreenSeeker and 27% for Crop Circle, from the first (7:30 am) to the last reading (11:30 am). Near infrared showed small differences in the readings, not significant, unlike the results of the visible data, but they both increased. These results lead to conclude that the visible wavelength is more affected by the presence of dew than the near infrared. Then, the presence of dew should be considered for nitrogen recommendation based on active optical sensors.

Stressed plants can show a decrease in the absorption by the chlorophyll, also decreasing the reflectance in the near infrared due to changes in the cells structure. And that is what makes the sensors to be promising while evaluating nitrogen stress. Many studies were realized to estimate nitrogen deficiency in corn, wheat, edible beans, cotton, citrus, barley and sugarcane. These authors showed great potential to use sensors to estimate nitrogen content in crops.

Use of Optical Sensors Combined with a GPS Receiver to Create Maps

Maybe one of the most interesting applications for optical sensors in agriculture is to be able to use geographical coordinates to create maps from reflectance measurements. Precision Agriculture is based on the fact that all fields have variability. Understanding this variability is first step to make decisions about investments in Precision Agriculture. Many procedures may be used to characterize and treat spatial variability on yield aiming profit for the farmers, but a wide and safe vision about the impact of the variability in a production system requires an accurate measurement of this variability.

Soil variability is caused by climate, topography, vegetation, soil forming processes and also management. These factors can influence the variability in different scales and cause great variability on nutrient availability in the soil. Then, when using uniform rates of fertilizers it is almost certain that excessive rates will be applied to some parts of the field and inadequate rates in others.

Precision Agriculture can be defined as a management system that considers spatial variability that is present in a production field, independently of the size and treats locally this variability. It is well proved that quality and yield are spatially variable and systems are being developed to explore these variations and increase profit. The variable rate application of fertilizers is one of the options to manage variability, and creating maps from optical sensors can help to realize variable applications with nitrogen.

An optical sensor connected to a GNSS (Global Navigation Satellite System) receiver is able to register reflectance values with a pair of coordinates (latitude and longitude), and when the file is imported to a GIS software (Geographic Information System) it is possible to represent the measurements and its distribution in a field with maps.

Depending on sensor system used, that information can be stored in different ways and types of files. Some systems installed in agricultural machinery allow using a laptop with software that will store the sensor readings and coordinates. Other systems use their own data logger or a pocket pc. The figure is an example of a raw data collected in a field. Each strip is around 24 meters from each other, but what looks like a line, actually are many dots very close to each other due the frequency used to collect the data in the field.

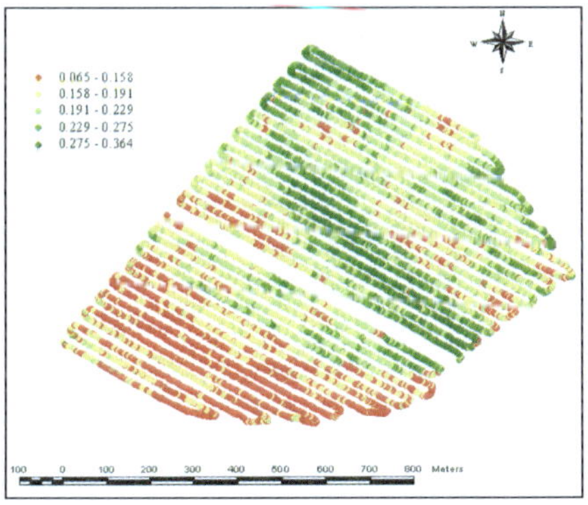

NDVI raw data

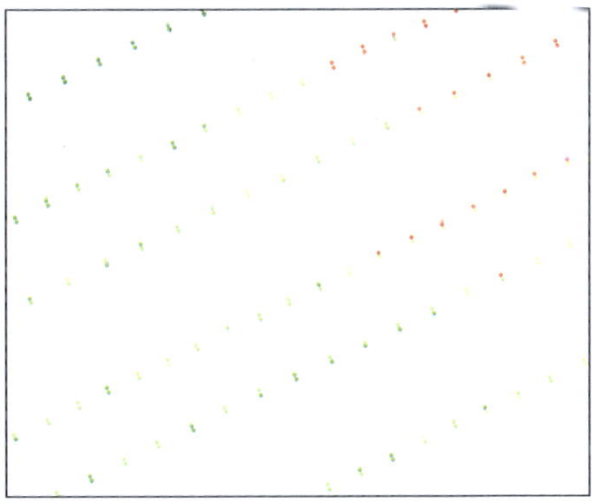
NDVI raw data (zoom in)

As showed before in the picture of the motorcycle with one sensor on each side, when looking closer at the data we can see that in every coordinate we have a NDVI value from each sensor. As the frequency is very high, lets say 5 Hz or 5 measurements per second, if the motorcycle moves at 5 m/s (18 km/h) with 24 m distant from the next pass, the sensor registers two NDVI values every meter and more than 800 measurements per hectare. But the raw data may have noise, so after applying some statistical filters to remove outliers and using some kind of interpolation method, for example the inverse distance, it is possible to create a surface map with a regular grid that will look like a raster image. The figure is a NDVI map after interpolating the raw data from a corn field.

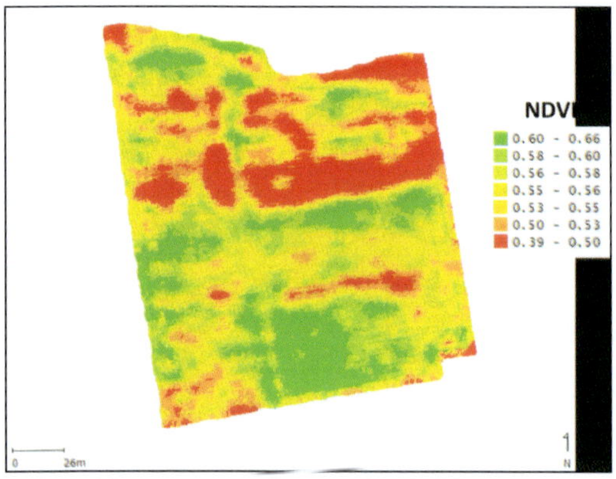

Interpolated NDVI map of a 2.86 ha field with a corn crop

Corn plants with low NDVI values

This maps shows that in the red areas the corn plants had a low development and because of that will have lower yields (less than 10 tons/ha), while the green areas had yields over 12 tons/ha. This information is important for precision agriculture users that are interested in finding the specific problem for that particular location. Looking at plants in the field, right in middle of the map in the red area, we can see the plants as showed in figure above.

The NDVI map created while the corn crop is still at the V6 stage of development, allow the farmers and agronomists to go back in the field and see what is causing this slower growth of the plants. And as the information is stored, it is possible to go back every year at the same spot for comparison, collect soil and plant samples, send to the lab and try to understand what the problem is. That means the farmer can treat each part of the field accordingly with its own characteristics.

Nitrogen Application based on Sensor's Measurements

Why manage nitrogen in agricultural crops? Because some crops do not have the ability to fix atmospheric nitrogen like soybeans do, so nitrogen is applied to crops from mineral fertilizers. Among the nutrients, nitrogen (N) is the most important and essential for crop development and is also important from an environmental perspective. Nitrogen is a constituent of chlorophyll, the first pigment to absorb the light energy necessary for photosynthesis. Plants usually have 1 to 5% nitrogen in their tissues, and if used appropriately with the other nutrients, N addition can accelerate the development of corn and other grains. And unlike other important nutrients as phosphorus or potassium that are measured in the soil with chemical analyses of soil samples, nitrogen is recommended based on historical response curves.

Plants with N deficiency have yellow leaves and reduced growth. Not only does the absence of N limit yield, but so can excess N. In most cases, N application recommendations are based on average conditions, and farmers do not use the most advantageous fertilizer combinations, even when there is no financial limitation.

The absorption of N by crops is variable among and between seasons, as well as between locations in the same field, even when the N supplies are high. The N supply from soil to crop varies spatially. Consequently, the demand for N by the crop also varies. As a result, the crop's nutritional status is a good indicator of the necessary N rate application.

Traditional N management around the world is generally inefficient, creates environmental contamination and is controversial. To provide appropriate recommendations of spatial N applications, it is necessary to use several tools simultaneously, such as crop and soil sensors, to achieve reliable measurements of N availability from soil and crops needs.

The evaluation of nitrogen use efficiency (NUE) in agriculture is an important way to evaluate the density of N applied and its role in yield. Because crop responses to N application depend on the organic matter in the soil, strategies of N management in cereal crops that include reliable predictions of the response index in each season could increase NUE.

In this scenario, sensors are becoming more prevalent in agricultural lands. Using variable rate equipment, it is possible to detect variability in crops and make rapid decisions in the field. Some sensors allow real time changes in agricultural practices by detecting variability and responding to that variability.

In reference, they developed a methodology to apply nitrogen based on crops reflectance, using a yield parameter sensible to local conditions, intrinsic and that could reflect yield potential, possible to be used in season. Unlike other models that need several parameters to predict plant growth, the optical sensors use the plant as an indicator. Later they created an index called INSEY (in-season estimated yield), which is calculated dividing the NDVI value by the number of days from sowing to sensing. The relation between the INSEY and Yield generate an exponential model using plot experiments, which can be used to estimate yield just based on the sensor measurements. Estimating yield where there is and where there is not nitrogen, by the difference in yield it is possible to calculate the nitrogen rate to make them to have the same yield.

This method considers the spatial variability of yield potential, the absorption of nitrogen applied at sowing and the crop's response to an additional N rate. NUE were increase in 15% and showed that the measurement of crop's reflectance can be used to calculate more efficient and profitable N rates. The ability to identify areas where the crop will respond to the fertilizer applied is important, if the response to N is expected, then management strategies can be modified.

Many field experiments were realized in order to create the algorithms that convert the sensor readings in nitrogen rates and to validate the results. Depending on the year, savings of none to 75% of nitrogen were found. For example the figure below shows the NDVI and yield response to nitrogen rates. Using the 120 kg of nitrogen per hectare as the standard rate, the NDVI measured 50 days before the harvest indicates that the wheat crop won't respond to rates over than 60 kg of nitrogen per hectare. And the wheat yield obtained after the harvest shows that rates over than 60 kg/ha will affect yield negatively. Therefore, in this example, the nitrogen rate was 50% lower than the commonly recommended rate.

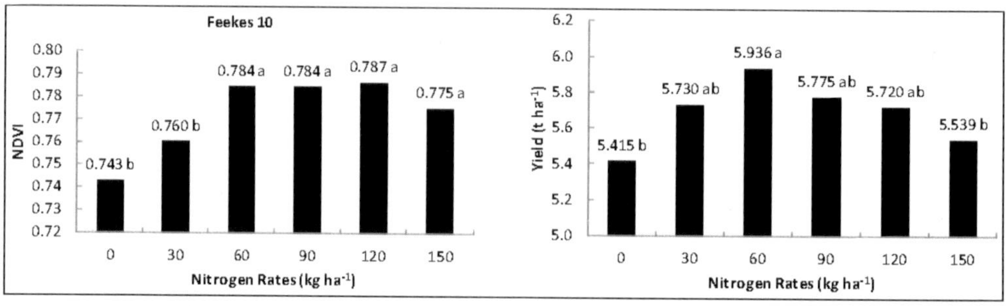

NDVI and wheat yield response with nitrogen rates applied

The yield goal, often used for nitrogen recommendations, is the yield that the farmer expects to produce. But what you expect to produce and what really is going to be produced are usually different. In they exemplify that the North Dakota State University recommend using the maximum yield obtained in the last 5 years as yield goal, because it is usually 30% higher than the average. The yield goal can also be calculated adding 5 to 10% to the average yield from the last 5 to 7 years. One example is to use 25 kg of nitrogen per ton of corn expected. But management practices and the weather have a huge influence on yield. Climate can vary significantly from one year to another, what may cause great differences in yield potential.

Now let's use an experiment as an example of how to realize a nitrogen recommendation using variable rate across a field. The experiment was realized in South Brazil, Paraná State, farm Manzanilha (25° 28' 55" S, 50° 21' 07" W), during 2007 winter. The sensor used was the GreenSeeker Hand Held measuring the NDVI. The treatments were disposed in strips with fixed rate of nitrogen (6 x 600 m) and variable rate of nitrogen (11 x 600 m), with four treatments and four replications. The wheat was sowed in May 28th. The Sketch is presented in the figure below.

Table: Treatments with nitrogen rates (kg/ha).

Treatments	NSowing	NFeekes 1	NFeekes 5	NFeekes 10	Total N
1	18,4	0	0	Sensor	18,4
2	18,4	34	0	Sensor	52,4
3	18,4	0	102	0	120
4	18,4	34	68	0	120

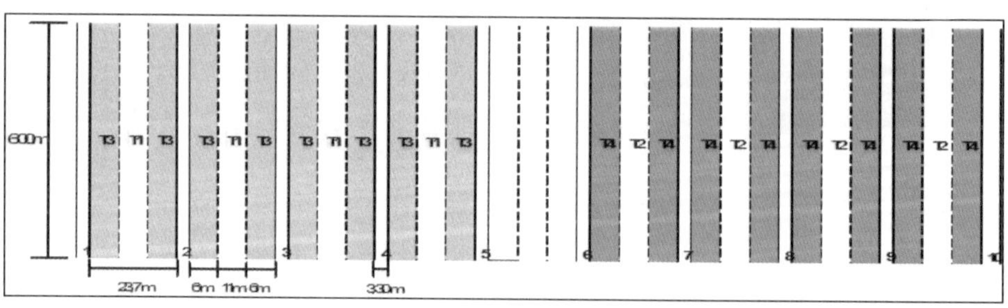

Experiment sketch

All treatments received 18.4 kg/ha of nitrogen at the sowing; treatments 2 and 4 had an extra 34 kg/ha just after sowing and treatments 3 and 4 received more 102 and 68 kg/ha of N, respectively, on Feekes 5 to complete the 120 kg/ha of N, recommended based on historical response curves. Treatments 3 and 4 were used as control, and treatments 1 and 2 received variable rates of nitrogen at Feekes 10 based on the sensor readings.

The measurements were realized with the GreenSeeker sensor mounted on a motorcycle 79 days after sowing. The pictures of the field and data collection can be seen in figure.

Details of the sensor mounted on the motorcycle (a) and data collection (b)

After the measurements, the data were imported to a GIS software (SSToolBox) to create a map with the NDVI raw data.

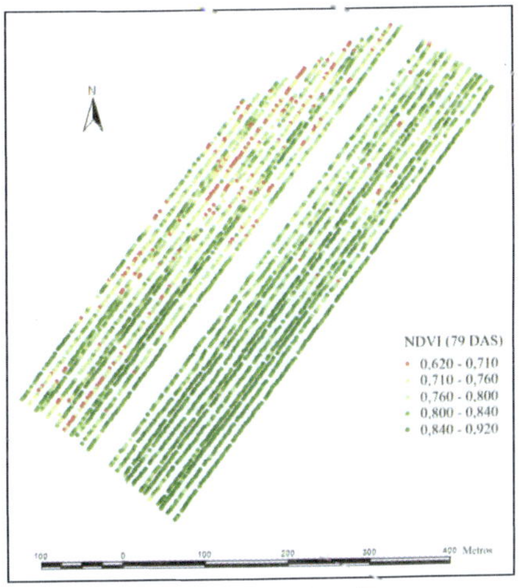

NDVI raw data in wheat

Looking at the NDVI map it is possible to notice that even at the 120 kg/ha of N strips there was spatial variability, so it was decided to adapt the methodology proposed in, and do not use the average of the strip as a control, but divide the strip in three levels according with the topography. The elevation map was created from the data collected by the GPS receiver. It was used three NDVI averages for treatment 3 and more three averages for treatment 4, with a total of six NDVI values to be used as a reference and estimate yield.

After dividing the elevation in three ranges it was calculated the average of the NDVI values for the 6 regions. Then the response index (RI) was calculated with the ratio between the average NDVI from strips with 120 kg/ha of N and the NDVI from every point in the map. That means that every point has an NDVI value and a RI value.

$$RI = NDVI_{\text{rich strip}} / NDVI_{\text{field}}$$

Which,

$NDVI_{\text{rich strip}}$ = NDVI from 120 kg/ha of N strips;

$NDVI_{field}$ = NDVI from all points in the field that will be applied N;

RI = response index.

The response index means that if the number is higher than 1, the NDVI from the strip with more N is higher than the NDVI from the parts of the field that will be applied, so the plants are using the N. And if RI = 1 means that where there is N and where there is no N the NDVI is the same, so there is no need to apply more N. Using RI it is possible to identify the parts of the field that will need more nitrogen, and save where there is no response, because it will not have increase in yield.

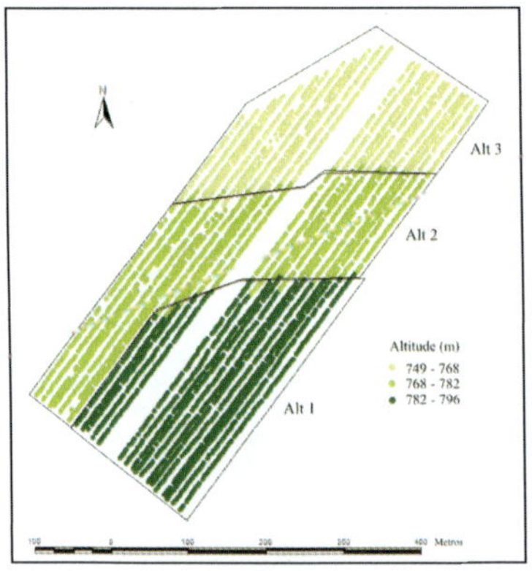

Elevation map divided in three levels

To calculate the N rate was used the methodology from, identifying the difference of yield between the N rich strip and the rest of the field, and then applying the correct N rate to reach the same yield. That means the objective is to save nitrogen and not increase yield. The N rates calculated were used to create a nitrogen application map.

The calculated rates varied from 0 to 60 kg/ha of N, but they were simplified into three ranges and applied the maximum of each range. For example, the rate 20 kg/ha represents the rates from 0 to 20 kg/ha and so on. This simplification was done to make the application easier. To apply the nitrogen it was used a sprayer mounted on a tractor for liquid nitrogen application. To vary the rates along the field the tractor changed the speed, for each one of the rates was determined a specific gear of the tractor and a specific engine rotation. The sprayer was calibrated to apply 20 kg/ha (2.5 m/s), 40 kg/ha (1.25 m/s) and 60 kg/ha (0.8 m/s). The gears and the engine rotation for tractor were determined with a 50 m space measuring the time that the tractor took to move the 50 m.

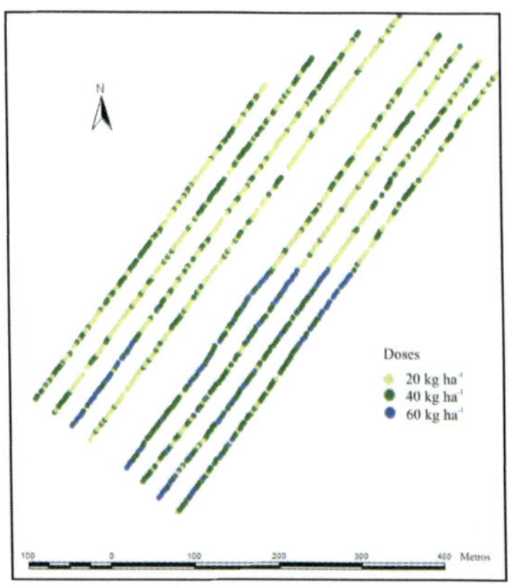

Nitrogen map from the calculated rates

Pictures of the liquid nitrogen fertilizer application

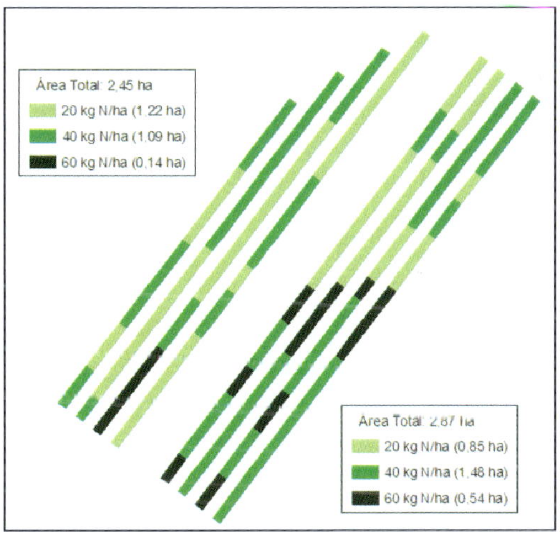

Sketch of the nitrogen application

The harvest of the strips was realized using a combine John Deere 9650 STS with a 30 feet header and equipped with a yield monitor AgLeader PF3000 Advantage, what made possible to generate the yield map. The above figure shows the yield map that was similar to the NDVI map. We can see in the map that each strip has a pair, because one strip is the 120 kg/ha of N and the other is the variable rate application. The yield varied from 1,500 to 5,500 kg/ha of wheat. But even the strips with 120 kg/ha of N had some parts of the field with lower yields, which means that these regions of low yield have some other problem and not nitrogen deficiency. Comparing with the application sketch the region of higher rates are same regions of high yields.

The left side of the experiment (first 4 pair of strips) had an average yield of 3,053 kg/ha using variable rate application, with 69.3% of saving in nitrogen, and 3,026 kg/ha using 120 kg/ha of N. The right side of the experiment had an average yield of 3,568 kg/ha using variable rate, with 42.5% of saving in nitrogen, and 3,546 kg/ha using 120 kg/ha of N. There was no statistical difference between the yield of fixed rate and variable rate, but the savings were high. The saving of nitrogen was higher in the region with lower yield potential.

As high is the yield goal just with the nitrogen supplied by the soil, with no additional N (low RI), in general, lower will be the N rates to reach maximum yields. If the RI for a field is low (RI < 1.1), that means the places with no N applied are similar with the places with N, and probably the response to an additional application will also be low. But if the RI is high (RI > 1.1), probably the crop will respond to the fertilization, so complementary rates should be applied.

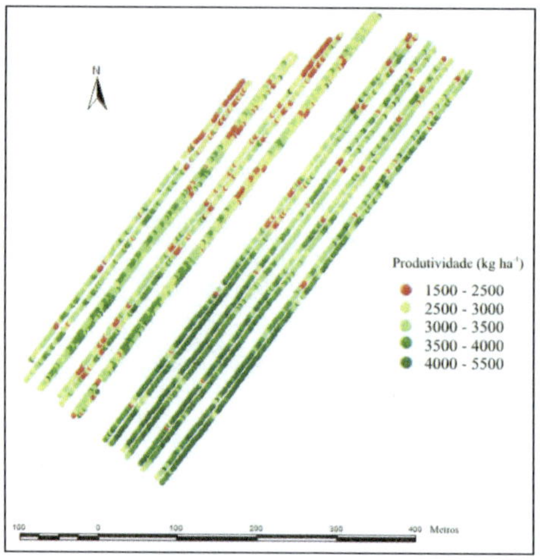

Yield map collected by the John Deere combine

Ps: In each pair of strips, the left one is the 120 kg/ha of N and the right one is the variable rate application.

There is NDVI spatial variability even in the regions where fixed rates of nitrogen were applied, showing that crops respond in different ways inside the same field. The methodology used to apply nitrogen in variable rate, using the crop as an indicator and optical sensors to measure is really promising, being able to save nitrogen in places where there is less use by the plants.

However, optical sensors do not consider yield potential. Reflected light by the crop in a high yield zone can show an appropriate content of N at the moment of measurement, but N can still be deficient before maturation. In a low yielding zone, reflected light may suggest a higher need for N, inducing an excess rate application for plants that are limited by other factors.

Large differences were found between two corn fields, one with yields <12,000 kg/ha and the other with yields >17,000 kg/ha. These authors suggested that considering climate conditions between planting date and the measurements could lead to a better relationship between yield estimation and actual yield because climate varies among areas and among years. Even when data from different areas and years are combined, the resulting model may not be reliable due spatial and temporal differences. These differences are some of the primary problems in developing a model that can be used widely to estimate yield by optical sensors. They also highlight that there is much room for improvement.

Variables that affect N availability were included, such as organic matter and soil exchange capacity, and variables that are related to N demand by the crop, such as solar radiation. Additionally, the variability of rain distribution and water stored in soil at the time of planting can affect yield and management decisions.

Based on the observations of long-term experiments, it was found that the treatments with higher yield did not always correspond to higher N application rates. In some years, maximum yields were observed with low N application rates. In 3 long-terms experiment, there was no relationship between RI and yield. These results indicated that RI and yield are independent, and so they need to be estimated separately. Any practice that does not consider the independence of RI and yield can lead to misinformed recommendations. If yield potential varies among years along with

N demand, the obvious solution to the increased efficiency of N recommendations is to be able to estimate crop response to N and yield potential.

Soil Moisture Sensors

The quantity of water in soil is called the soil moisture content. After rainfall or irrigation, some water drains from the soil by the force of gravity. The remaining water is held in the soil by a complex force known as surface tension and varies depending on the amount of sand, silt, and clay. Sands, with larger particles and smaller total surface area, will hold less water than clays, which have much smaller particles and larger total surface area. The drier the soil, the greater the surface tension, and the more energy it will take for a plant to extract water.

Vineyard managers often measure soil water content as a guide to determine their irrigation timings and amounts. There are several methods for monitoring soil water content. Correlating these methods with actual inches of moisture per foot of soil is very complicated but at the very least can help a grower to identify patterns of water use, depth of irrigation, and soil water content trends over time.

Tensiometers

A tensiometer, as its name implies, is a device for measuring soil moisture tension. The design is a simple tube with a porous cup at the lower end and a vacuum gauge on top. The tube is filled with water, sealed airtight, and placed in soil. As soil dries, water is pulled from the porous cup into the soil, creating a vacuum and causing the gauge to move. As soil continues to dry, more water is pulled out and the suction increases. As soil re-wets after a rain or irrigation, water moves back into the cup and the suction decreases. Installing tensiometers in soil requires attention to detail to obtain accurate readings.

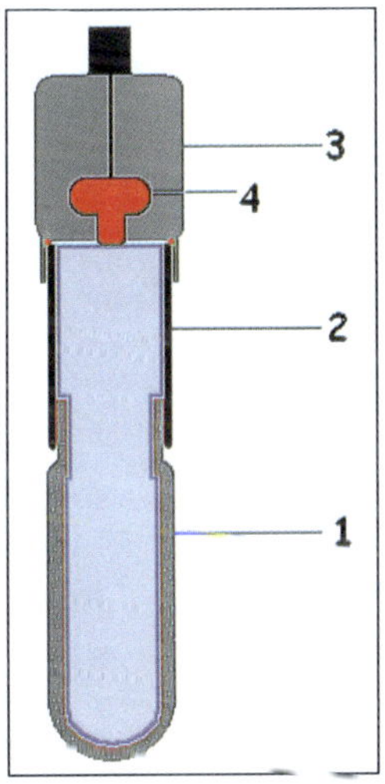

Electronic tensiometer: (1) Porous cup; (2) Water-filled tube; (3) Sensor-head; (4) Pressure sensor

Tensiometers are usually placed as a pair with the shorter tube positioned in the middle of the rooting zone (e.g., 18 inches deep) and a longer tube positioned near the bottom of the rooting zone

(3 to 4 feet deep). Growers can use the difference between the two tubes to monitor water usage and determine the effective depth of irrigation. At least two stations (two tubes per station) are recommended per field, or more depending on soil variability.

Tensiometers have the advantage of being inexpensive, and easy to install, maintain, and read. They are better in fine-textured soils where good contact can be made between the porous cup and the soil. They do not work well in coarse sands where good contact may not be possible. Because the gauges are aboveground, the units are prone to damage by vineyard equipment.

Electrical Resistance Blocks

Electrical resistance blocks are also known as gypsum blocks or soil moisture blocks. They are simple devices with two electrodes embedded in a block of gypsum or other similar material. When blocks are buried in soil, water moves into or out of the block, depending on the moisture of the soil, changing the resistance between the two electrodes. Like tensiometers, gypsum blocks are cheap and easy to install. They are usually installed in at least two stations per field, at two depths, and must be installed correctly to provide accurate readings. Some block designs perform better under wet soil conditions and some correct for soil temperature. The meter used to read the blocks can be moved from field to field, but is specific to the block design (i.e., it is not a simple ohm meter). The wires aboveground are much less prone to damage by equipment compared to tensiometers.

Neutron Probe

A neutron probe uses a radioactive source for measuring soil moisture. A tube, usually made of PVC or aluminum, is installed in soil to a depth of interest and the radioactive probe is lowered into soil to measure soil moisture at as many depths as desired. The probe emits fast neutrons that are slowed by water in the soil in a way that can be calibrated to the soil water content. The probe has a significant advantage, especially for perennial crops, because access tubes are easy to install and relatively permanent. Another advantage is the reading accounts for a spherical area about 10 inches in diameter, much greater than other methods. The major limitation to this method is the probe itself; it is expensive and the presence of a radioactive source triggers requirements for operators to be trained and licensed in handling, storage, and use. In some production regions, service providers are available, usually at a fixed cost per access tube for a growing season.

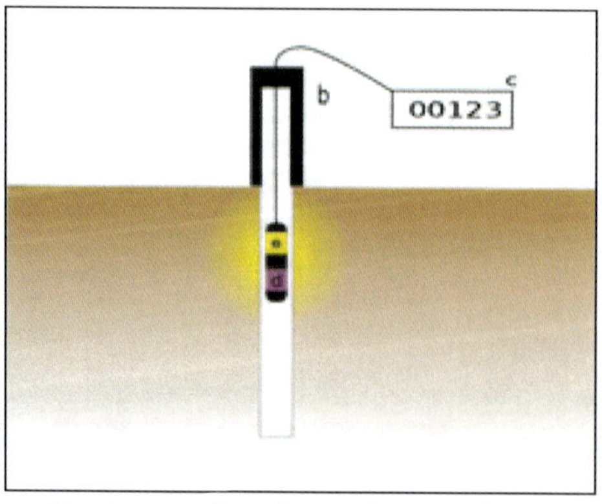

Neutron probe schematic: e=emitter; d=detector; b=shield; c=counter

Di-electric Sensors

Di-electric sensors measure the di-electric constant of soil, a characteristic that changes with changing soil moisture. A common method is called time domain reflectometry, or TDR. The advantage of these types of systems is that they are designed to be left in place and provide continuous readings of soil moisture. The disadvantages are that the units are expensive and read soil moisture only a very small distance from the unit.

Dielectric Soil Moisture Sensors

Dielectric soil moisture sensors determine the soil moisture by measuring the dielectric constant of the soil, an electrical property that is highly dependent on the moisture content. The constant for a dry soil is between 3 and 5; about one for air; and is about 80 for water. Thus, changes in the moisture content cause a substantial change in the dielectric constant of the soil. Calibration equations correlate the dielectric constant with soil moisture content. The most common dielectric devices are capacitance sensors and time-domain-reflectometry (TDR) sensors although other types of dielectric sensors exist. Virtually all of the sensors listed below can be attached to data loggers for a continuous record of soil moisture content. Many of these systems have the option of radio or cell phone telemetry that can link the sensors directly to an office computer or be viewed through a web-based server.

Capacitance Sensors

The capacitance sensor consists of two electrodes separated by a material called the dielectric, a material that does not readily conduct an electrical current. Normally, cylindrical-shaped electrodes are used. Inserting the sensor into the soil results in the soil becoming part of the dielectric. An oscillator applies a frequency between 50 and 150 Mhz (depending on the manufacturer) to the electrodes, which, in turn, results in a resonant frequency, the magnitude of which depends on the dielectric constant of the soil. The greater the soil moisture content, the smaller the frequency. The frequency is then used with a calibration equation to estimate volumetric soil moisture content.

Capacitance instrument

Time-Domain-Reflectometry (TDR)

Time-domain-reflectometry involves installing two or three steel rods, called waveguides, into the soil parallel to each other. Components of a TDR system are a voltage pulse generator, a signal analyzer, the waveguides, and a cable connecting the waveguides to the instrumentation. An electrical pulse applied to the waveguides travels their length, and then is reflected back when it reaches the end of the waveguides. The travel time required for the pulse to reach the end of the waveguides and back depends on the dielectric constant of the soil. The larger the dielectric constant, the longer the pulse travel time.

TDR probe

Calibration

Calibration equations relating the dielectric constant to the soil moisture content are necessary. These equations generally are provided by the manufacturer; however, in some cases, site-specific calibration may be needed. Some manufacturers provide a universal calibration, while others provide calibrations based on soil texture. Some calibration curves provide the volumetric soil moisture content. Others may display the percent of available soil moisture remaining in the soil.

Field calibration may be necessary in fine-textured soils of California even though a manufacturer calibration curve is provided. Field experience has shown that unrealistically high readings can occur in these soils. If the sensor reading just after an irrigation (when the soil is at field capacity) exceeds 50% (6 inches/foot), then the sensor should be calibrated for the site-specific conditions. To determine if the dielectric sensor reading is realistic, compare the field capacity volumetric soil moisture content with the sensor value. If the values differ greatly, then a site-specific calibration is needed. Also, in some cases, very low sensor readings (approaching zero) have been found just before irrigation, indicating a need for calibration.

Developing a calibration curve involves collecting soil samples very close to the sensor and determining their volumetric soil moisture content and making a sensor reading at that depth at the same time.

Installation

All dielectric soil moisture sensors must be carefully installed to prevent air gaps between sensor and soil. Air gaps cause errors in their readings. Some sensors require specialized equipment for installation, while others are installed by placing the sensor into a pilot hole. In some cases, slurry is used to obtain a tight seal between sensor and soil.

Zone of Influence

The zone of influence is very small compared to the neutron moisture meter. The zone of influence is about 4 inches from the sensor for capacitance instruments and about 1 inch for TDR sensors.

Advantages

Advantages of dielectric soil moisture sensors include the ability to be left in place to continuously log soil moisture content; repeatability of measurements; sensitivity to small changes in soil moisture content; and their precise resolution with depth because of the narrow vertical zone of influence.

Disadvantages

Disadvantages include the need for a calibration equation; the difficulty in developing the equation; the relatively small zone of influence; possible influence of soil salinity on probe reading; and sensitivity to air gaps surrounding the sensor.

Capacitance Sensors

Capacitance sensors measure the dielectric constant of the soil in order to find its water content. Since the dielectric constant of water is much higher than that of air or soil minerals, the dielectric constant of the soil is a sensitive measure of water content. Because of the difference in the dielectric constant between water and air and the very small amount of soil the sensors evaluates, any air gap around the sensor will cause large errors. Always closely follow the manufactures installation instructions to minimize this problem.

The sensors come in a variety of configurations, but basically consist of a pair of electrodes (either a group of parallel spikes or circular metal rings) which form the capacitor with the soil acting as the dielectric in between. This capacitor works with an oscillator to generate an AC field. Changes in soil water content are detected by analyzing the results with the electronics.

The advantage of the capacitance sensors is that they allow for a great deal of flexibility in their design configuration and the cost of the electronic components is relatively low. Because of this, they are the most common fully electronic soil water sensors on the market today. Several companies produce systems, and they all look quite different. Some models are designed to be installed at one location for the entire season, and other models are designed to be portable. The portable models are generally of two types: one being designed with electrodes that are pushed into the soil and the other operates from within an access tube that needs to be installed in advance. Another advantage is that most systems come with a factory calibration that can be used in all soil types, even though developing a calibration for the soil the sensor will be used in may improve the accuracy some.

The disadvantage of the capacitance sensors is that they need to be very precisely installed to avoid any air gaps between the sensor and the soil. The systems are in the upper end of the cost range for sensors used in production agriculture and some researchers have reported observing accuracy problems.

Watermark Sensors

Granular matrix sensors, commonly know as Watermark Soil Moisture Sensors reduce the problems inherent in gypsum blocks by use of a granular matrix enclosed in a metal and plastic capsule. It may be used in all soil types, but is limited to a range of 0-240 centibars, which works well for most irrigated crops. The sensors operate on the same electrical resistance principle as gypsum blocks and require a different meter that is specially designed to measure the electrical resistance of the Watermark sensors. The meter is calibrated to read in centibars, the same units used with tensiometers, which in turn reflects the soil water potential.

Installation and operation of the Watermark sensors are similar to soil moisture blocks. The Watermark is more expensive than the soil moisture blocks, but by gluing on a length of PVC pipe, it can be easily retrieved at the end of each growing season and reused for several years.

Watermark sensors have the advantage of being easy to install, low cost, reusable, and it is easy to add extension wires to run several hundred feet across a field to a data logger or a readout point located at the edge of the field.

References

- Remote-sensing-applications-in-agriculture, remote-sensing: grindgis.com, Retrieved 25 Februray, 2019
- Advantages-and-uses-of-Agriculture-Sensors, Terminology: rfwireless-world.com, Retrieved 26 July, 2019
- Sensors-agriculture-smart-intelligent, frost-perspectives: frost.com, Retrieved 21 may, 2019
- Sensing: cropwatch.unl.edu, Retrieved 8 january, 2019
- Optical-sensors-applied-in-agricultural-crops, optical-sensors-new-developments-and-practical-applications: intechopen.com, 1 Februray, 2019
- Soil-moisture-sensors: extension.org, Retrieved 13 May, 2019
- Drought: ucmanagedrought.ucdavis.edu, Retrieved 6 January, 2019

4

Irrigation Technology in Agriculture

The application of controlled amounts of water to plants when needed is known as irrigation. It helps in the growth of agricultural crops, revegetating disturbed soils and maintaining landscapes in areas that are dry during periods of less rainfall. All the diverse methods of irrigation as well as the technologies related to it have been carefully analyzed in this chapter.

Technology can, and already does, aid agriculture in innumerable ways. One prominent part of agriculture that can use technological innovation to increase efficiency and effectiveness is irrigation.

As the world's population continues to increase at a fast pace, more food and water will be needed to sustain humanity. In the past 50 years, we have tripled our need for water and food, and there are no signs of this trend slowing down. As a result of these conditions, smart, innovative agricultural practices are needed now more than ever. Technology can, and already does, aid agriculture in innumerable ways. One prominent part of agriculture that can use technological innovation to increase efficiency and effectiveness is irrigation.

Irrigation is the method in which a controlled amount of water is supplied to plants at regular intervals for agriculture. It is used to assist in the growing of agricultural crops, maintenance of landscapes, and re-vegetation of disturbed soils in dry areas and during periods of inadequate rainfall. Additionally, irrigation also has a few other uses in crop production, which include protecting plants against frost, suppressing weed growth in grain fields and preventing soil consolidation. Irrigation systems are also used for dust suppression, disposal of sewage, and in mining. Irrigation is often studied together with drainage, which is the natural or artificial removal of surface and sub-surface water from a given area. Irrigation has been a central feature of agriculture for over 5,000 years and is the product of many cultures. Historically, it was the basis for economies and societies across the globe, from Asia to the Southwestern United States.

An agricultural irrigation system

Agricultural irrigation accounts for roughly 80% of the ground and surface water used in the United States, and could account for up to 90% of water used in western states. Some 53.5 billion gallons of groundwater are used daily for agricultural irrigation from 475,796 wells. Irrigated agriculture is an indispensable part of the nation's economy as it contributes significantly to the value of U.S. agricultural crop production, as well as supporting livestock and poultry production through irrigated animal forage and feed crops. However, all of this freshwater use in irrigation is wasteful. Worldwide, agriculture accounts for 70% of all water consumption, compared to 20% for industry and 10% for domestic use. Roughly 15-35% of irrigation withdrawals are estimated to be unsustainable, and the agriculture industry wastes 60% or 396 trillion gallons, of the 660 trillion gallons of water it uses each year.

Water Measurement

The key to a sustainable agriculture looking to the future is in water use, and it is clear that the current standards in place are much too wasteful to rely on. Water metering and measuring technology is at the heart of this issue, but there is not only one method or technology used to measure water flow in irrigation systems. In irrigation systems, water used to irrigate land is carried under pressure to its destination via pipes. Flow occurs in a pipeline when a pressure difference exists between the two ends of the pipe. The rate or discharge that occurs depends mainly on the difference in pressure that exists from the inlet of the pipe to the outlet, the friction or resistance to flow caused by the pipe's length, pipe roughness, bends, restrictions, changes in pipe shape and size, and the cross-sectional area of the pipe. Most flow measurement devices measure flow, or discharge, indirectly. These devices arc commonly classified into 2 types – those that measure velocity and those that measure pressure. Regardless of the type of device, the pressure or velocity is measured, and then charts, tables, or equations are used to obtain the discharge, or flow. Some water measuring devices that use measurement of pressure to determine discharge include weirs, flumes, orifices, and Venturi meters. Other devices that work differently from those that measure water pressure include acoustic water meters, magnetic water meters, pitot tubes, and rotameters. All of these non-pressure-based devices work differently to determine water discharge.

In some acoustic or ultrasonic meters such as transit time meters, the velocity of sound pulses in the direction of water flow is compared to the velocity of sound pulses opposite to the direction of flow to determine the mean velocity and, thus, discharge. With Doppler acoustic meters, sound pulses are reflected from moving particles within the water mass, like radar. Pitot tubes measure water flow by relating velocity pressure, $V^2/2g$, to discharge, and rotameters do via flow indicators suspended in tapered sections of pipe. Lastly, in magnetic meters, flowing water acts like a moving electrical conductor passing through a magnetic field to produce a voltage that is proportional to discharge.

Ideally, a water meter is accurate, reliable, efficient, and compatible with the rest of the irrigation system it is part of. Accuracy is an obvious characteristic. Water needs to be accurately metered to avoid waste. Water meters need to be reliable for a similar reason – if the meter is breaking down frequently, there is no accurate way to measure water usage, which will result in waste. Water meters need to be efficient in terms of their installation and operation. Special requirements for installation only result in wasted time during the installation process, which, in a general sense, results in more wasted. Lastly, a good water meter is compatible with its irrigation or crop control system. Todays integrated irrigation and crop control systems vary in communication requirements. Using a meter that is compatible with data logging and telemetry software is hugely important in terms of the meter's accuracy, operation, and maintenance.

Waterflux 3070 Meter

Magnetic meters are innovating more than others, and are more completely meeting the requirements for an ideal water meter. One magnetic meter in particular that is perfect for irrigation applications in agriculture is WATERFLUX 3070 magnetic meter. The WATERFLUX 3070 is a battery powered all-in-one (integrated flow, pressure, and temperature measurements) electromagnetic water meter. Unlike mechanical water meters, the WATERFLUX 3070 is maintenance-free and offers a much larger turn down ratio (1000:1). Its measuring tube with a rectangular and reduced cross-section enables a stable measurement even at low flow rates. Due to the optimized flow profile, the WATERFLUX 3070 can be installed virtually anywhere without straight inlet or outlet runs – behind pipe bends, slide valves or a reduction in the pipe. Even burial installation or use in flooded areas is possible.

The WATERFLUX 3070 has several key advantages for agricultural applications due to its rectangular reduced bore design. This unique reduced bore design acts as its own internal flow straightener, and also results in the meter creating a smaller magnetic field which has less power consumption, extending battery power life to up to 15 years with a KROHNE dual battery pack. The rectangular bore of the WATERFLUX 3070 is also effective in that it contracts flow profile disturbances, has an ideal coil arrangement for a strong and homogenous magnetic field, results in a doubled mean flow velocity, and enables a measurement that is stable and independent of the flow profile.

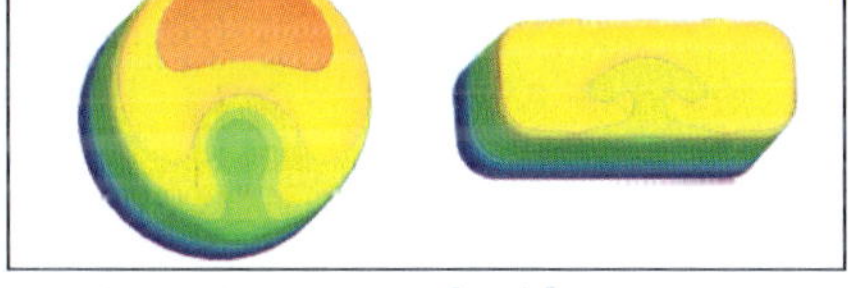

Flow velocity through pipes under ideal conditions (top row) and disturbed flow conditions (bottom row)

In all, the WATERFLUX 3070 is:

- Accurate: It features bi-directional flow measurement over a wide dynamic range and its unique rectangular sensor design gives good low-flow performance.

- Reliable: The WATERFLUX 3070 features a robust construction; with its sensor tube coated in Rilsan to guarantee a long life and its connection box is made from stainless steel. In addition, the WATERFLUX 3070 features no internal moving parts, making it virtually maintenance-free.
- Efficient: Installing the WATERFLUX 3070 is quick and easy. It has no inlet or outlet installation requirements, features plug-and-play connectors, and can be installed virtually anywhere.
- Technologically compatible: The WATERFLUX 3070 is fully data logging compatible and features multiple power and converter options.

To solve the water-use problems currently affecting the agricultural industry, better water metering technology is a necessity. Among the many types of water metering devices, innovative magnetic metering devices like WATERFLUX 3070 check all the boxes for ideal water metering technology, and are showing the way towards more efficient and effective water use in agricultural irrigation applications.

Micro or Trickle Irrigation Systems

The terms microirrigation and trickle irrigation can be used interchangeably. Trickle or drip irrigation were terms used to describe individual emitters or rowcrop emitters. Microirrigation is a broader term that is used to cover all forms of small emission devices including individual emitters, rowcrop tubing, spray strakes, and micro-sprinklers. All of these devices were developed for water distribution to small crop area, frequently to individual plants.

Microirrigation systems have very small openings, called emitters, that are designed to discharge the water at a known rate for the operating pressure. Different emitters have different sensitivities to pressure changes and water quality. Water filtration of 150 to 200 mesh is required to ensure good operation. Most emitters operate at pressures of 8 to 20 psi, with flow rates of 0.5 to 2.0 gallons per minute for individual emitters or 0.5 to 1.5 gallons per minute per 100 feet of length for rowcrop tubing.

Emitters are designed to be self-cleaning or to be taken apart for cleaning. Periodic chlorine injections are used to keep the system free of algae or bacterial slime. Acid injections, along with periodic system flushes, help to remove mineral buildup. Backflow prevention devices should be used to protect the water supply when chemical are injected.

Microirrigation irrigation systems require more planning than overhead sprinkler systems because the emitter systems deliver water directly to the plant root system. The crop influences the selection of the emitter system. There are many types of emitters to meet specific needs.

Water conservation, energy conservation, and automation are accomplished by micro-irrigation. Both water and nutrients can be delivered efficiently. Applications of the system can be found in all kinds of nursery/greenhouse production systems.

Liners can be watered using trickle rowcrop tubing or individual trickle emitters inserted in polyethylene pipe. Container-grown crops use rowcrop tubing, individual emitters in pipe, spaghetti tube systems, or spray stakes to lessen water requirements and achieve adequate coverage. On in ground or raised bed plantings, trickle rowcrop tubing can adequately wet the soil.

Drip Irrigation

Over recent years, there have been considerable technological advancements in irrigation. One of the more effective of which is drip irrigation. Irrigation technologies, put simply, provide water to plants, and methods for doing so can vary widely. Techniques for irrigation can range from surface irrigation techniques, either through channels or completely flooding the field, to the more precise and controlled method of drip irrigation. Other examples include overhead irrigation which, consequently, creates a lot of runoff.

Irrigation was a very early technological advancement of our species. It allowed for the development of more efficient farming and subsequently provided stable, more or less, food supplies. Drip irrigation is, in effect, a modern "tweak" of the time old technique.

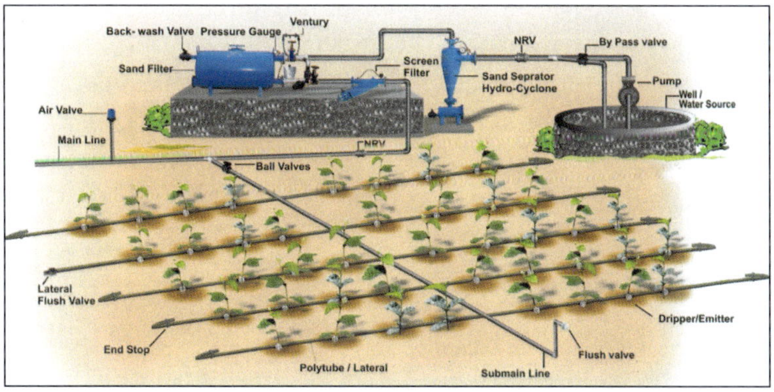

Drip irrigation is renowned for being a very efficient method of watering plants. By way of example, the average sprinkler system has an efficiency of around 75-85%. Drip irrigation, in contrast, has efficiencies in excess of 90%. Over time this difference in water delivery efficiency will make a considerable difference in crop production and the company's bottom line. In areas where water is short supply, such as the desert regions of the US, drip irrigation has, unsurprisingly, become the preferred method of irrigation. Drip irrigation systems are relatively inexpensive and easy to install, simple to design, and help maximize plant health due to the reduced moisture levels on fields.

This form of irrigation, sometimes called trickle irrigation, applies water directly and slowly to the soil. The technique's efficiency is provided by two major factors. The first is that water is absorbed by the soil for access to the plant roots rather than running off or evaporating. Secondly, the water is only supplied to those areas of the field that actually need the water, i.e. the plant roots. Most drip irrigation systems are simple to design which minimizes design errors and installation flaws.

Importance of Irrigation

Irrigation is one of the oldest technologies mankind developed. It is used extensively across the world. Countries with the largest populations (USA, China, India etc.) have over 100,000 km^2 of irrigated land.

Irrigation consumes a lot of fresh water and can result in water logging of crops and buildup of salts. Salinization is a big problem in places like Egypt. The riverbed of the Nile has been irrigated from close to 5000 years since around 3100 BCE. These practices draw salt from lower horizons in the soil to upper levels. This is so bad in some places that the soil is actually whitish in areas! This isn't an issue isolated to Egypt and occurs wherever irrigation has been employed for a long period of time.

Drip irrigation offers a great solution to this potential problem. Historic practices such as centre-pivot irrigation are not sustainable in the long term. They consume large amounts of water and are potentially damaging to the "health" of the soil. Drip irrigation allows the user to better control the amount of water plants receive, rather than blanket watering the area. Eutrophication is massively decreased by drip irrigation as fertilizers are not carried away by water runoff into watercourses.

Drip Irrigation might be the Future

Italy is one of the world's largest agrarian countries and has a large percentage of land given over to growing wheat, corn, rice and fruits etc. Italy began implementing drip irrigation in 2011. Italy's adoption of drip irrigation is estimated to save the country 4.3 billion Euros over the next thirty years. According to the World Water Development Report (WWDR), 47% of the world's populations are likely going to live in "areas of high water stress" by 2030. If this warning is to be believed, it is essential that we develop and implement ways to better use and conserve water supplies. Drip irrigation might just be the perfect solution for farming.

Working of Drip Irrigation

In effect, drip irrigation places small drip emitters in close proximity to the crops' root systems. This provides a much-improved efficiency and makes the system much more controllable compared to other methods. The emitters release water in a slow and steady fashion. The emitters are very small, about the size of a US quarter and are arranged in an array in the ground. These emitters are directly connected to the water source by feeder hoses. Another setup is to have the emitters built into the feeder hose instead of rows of independent emitters. This is called a trickle hose.

Invention of Drip Irrigation

The invention of drip irrigation is often attributed to one Simcha Blass. Simcha was an Israeli engineer and inventor who lived between 1897 and 1982. Simcha was an important figure in water development in Israel and he, with his son, initiated, introduced and developed drip irrigation systems.

Drip irrigation was tested in a primitive form in the 1920's but the modern technology as we know it was properly developed by Simcha in the 1930's in Israel. Its discovery seems to have been something of an accident. Blass, whilst spending some time in the desert regions of Southern Israel noticed something strange. He noticed that one tree near his location was performing much better than all other vegetation nearby.

When Blass went for a closer look, he noticed that a water pipe near the tree had a small leak supplying its root system with its very own regular slow supply. This accidental discovery led Blass to embark on a journey of trial and error testing various materials and water pressures for an ideal solution. It wasn't until the 1950's with modern plastics that Blass could take his technology to the next level. In the 1960's Blass was able to finalize the technology and patent the design.

"Gubbins" of a Drip Irrigation System

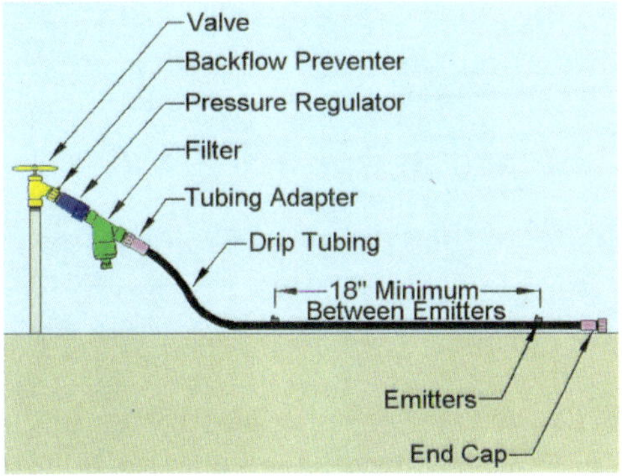

Simplified drip irrigation system

Drip irrigation systems are pretty simple setups but do consist of several constituent parts. A typical simple system will consist of the following components:

Valves

The role of valves in the drip irrigation system is very simple. They turn the water flow on or off. Valves come in various "flavors". Isolation valves are manually operated for systems that require an infrequent shut-off of the water. These valves are typically sited close to the water supply to enable isolating the system for repairs or off season. These can be installed anywhere in the system to enable isolation of segments of the system for localized repairs but this is usually only employed in larger systems.

Control valves are valves that turn the water on and off to individual "circuits" or areas of the yard that are perhaps irrigated separately from one another. These can be automatic (using solenoids) or manually operated. Depending on the system design there can be just one or many installed. For instance, you may have one control valve that controls water supply to the emitters in a vegetable garden. Yet another may be present that controls water supply to shrubs or hanging pots around the house and patio.

Drip irrigation system

Backflow Preventer

This is a piece of kit employed within the system to prevent, hence the name, dirt, bacteria, other contaminants from being sucked back up into the drinking water supply for the drip system. This device is essential for all drip irrigation systems.

Backflow preventers are essential because drip emitters rest directly on the soil and are potential very susceptible to water contamination from soil born diseases etc.

Pressure Regulators and Pressure Reducing Valves

These devices, as the name implies, reduce the pressure of water flowing through the system and keep it at a constant level. Pressure reducing valves and pressure regulators are in this instance synonyms and essentially the same thing.

Drip irrigation systems, on the whole, perform best at lower water pressures than typical water supply systems. These devices also enable a constant system pressure even if the supply pressure fluctuates periodically, which is nice. Designers need to be conscious of areas with low water pressure as these devices will clearly further reduce the system pressure.

Usually, two types of pressure regulators are employed in drip irrigation systems. Non-adjustable ones with pre-set outlet pressures and user adjustable types. In general, small homeowner system will use non-adjustable valves if they have less than 3 control valves. You can, of course, install adjustable valves if you want full control of your system. Non-adjustable regulators must be installed after the control valve and in cases where multiple control valves are present, pressure regulators are needed for each one. Accidental installation before the control valves can cause pressure surges that will damage the system.

Adjustable pressure regulators, on the other hand, can be installed before or after the control valves. In large systems, you can install a single or fewer adjustable pressure regulators in the main supply line before control valves to save on costs.

Filters

The filter is obviously used to, well, filter the water. Drip emitters have very small openings which are easily clogged up so employing filters earlier in the system is essential for increasing the life span of the irrigation system. Recommendations for filters are that they are between 150 and 200 mesh. High-quality filters are often installed before the valves or pressure regulator but lower quality ones can be installed after the pressure regulator. High-quality filters usually have a maximum pressure rating of 10.3 bars (150PSI).

Emitters

Now we come to the "guts" of the drip irrigation system. The emitters are responsible for directly controlling the rate of water supply to the soil. Emitters are usually small plastic devices that either screw or snap onto the drip tube or pipe. In trickle pipe systems they are pre-assembled and part of the pipe assembly. Common emitters, emit, water at around 4 liters per hour.

As a general rule of thumb 1 or 2 emitters are usually required per plant. This does, of course, depend entirely on the size of the plant in question. Trees or shrubs will clearly need more than a small plant. Use of multiple emitters also provides the system with backups in case of a blockage in one or more of the emitters. The more emitters present the wider the area of irrigation and hence the increased growth of roots for healthier crops and plants. Of course, if plants tend to be planted close together the system may only need 1 per plant depending on the system design and "coverage" of the emitters.

Emitters are usually installed at least 450 mm apart. As a rule, some sources suggest installing emitters 600 mm apart under 80% of the plant's leaf canopy; this is where the roots are after all. For highly permeable soils emitters should be placed 300 to 450 mm apart. Emitters should never be buried unless they are specifically designed for this purpose.

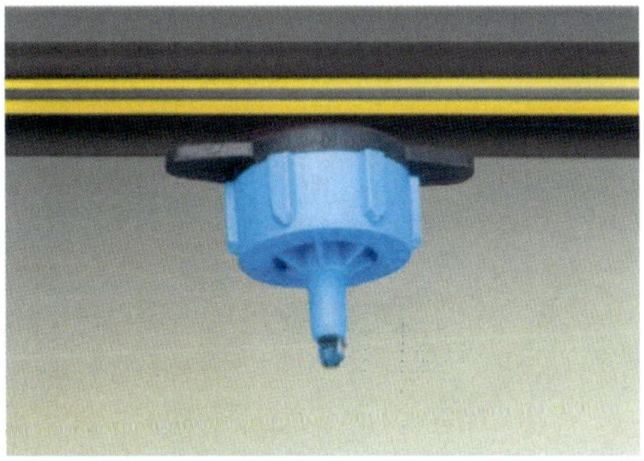

Mainline and Lateral/Sub-main Pipes

This pipe is the main connection between the water supply to the control valves of the drip irrigation system. It can be made of galvanized steel, copper, PVC or heavy wall Polyethylene. Each type has the inherent limitations and strengths. PVC, for example, is easily damaged by sunlight and is usually buried or protected. Polyethylene has a low burst pressure and is usually only used where water pressures are lower than 50 PSI.

Lateral/sub-main pipes are located between the control valve and drip emitter assemblies. These can again be made of PVC, PEX or Polyethylene. As these are generally placed downstream of the pressure regulator high-pressure ratings are not essential.

Drip Tube or Hose

This is a special type of tube common in most drip systems. They tend to be laid on the ground surface between plants. Emitters are generally installed to these tubes. Drip tubes tend to be made of thin-walled polyethylene and consequently have a much lower pressure rating than other parts of the system. It is generally recommended that these stay above ground as they can commonly be nibbled on by pesky local rodents. In large commercial setups, these tubes are usually "hard-piped" in these systems and the emitters are installed directly onto the laterals.

Drip tube tends to not exceed 60 meters in length from the point where water enters the tube. Tubes can be extended so long as the point of entry of the water supply never exceeds 60 meters from entry to pipe termination point. E.g. a 120-meter tube where the water entry point is at the center point.

Air Vent

Air vents are installed in systems that are turned off at any time. They prevent air from being sucked up into the emitters. As the water pressure falls away, air can be sucked back through the emitters and entrain dirt or soil into them. Clearly not desirable. The presence of an air vent mitigates this problem by drawing in air through it rather than the more delicate emitter openings.

End Cap or Flush Valve

Unless you want water to run out of the end of the drip tube you'll need to install an end cap. All well and good but this do introduce another issue for the drip irrigation system. The water flow within the drip system is very slow which can allow sediment to build up, even allow algae to grow within the pipes. Normally drip tubes are flushed about once a year, more if an algae problem is persistent.

Benefits of Drip Irrigation

Given the technology's setup, the greatest benefit this method provides to the producer is control. Given the amount of control it provides, the technique offers great economic benefits as well as reduced waste. A typical lawn sprinkler will use between 4 and 20 liters of water per minute. A standard drip irrigation system, on the other hand, measures water use in liters per hour. This slower supply of water to the crops improves root take up and reduces water loss through ground percolation. This allows the water to be used more efficiently and reduce waste through evaporation for example. The direct application of water to the soil also prevents drift. Drift is the phenomena of water being blown or dispersed to other parts of the site where water is not required, e.g. walkways etc.

A well maintained and managed drip irrigation system can, all but eliminate, water waste through surface runoff. Drip irrigation systems rarely need excavations and rarely disrupt the integrity of landscapes during installation. Tubing can be weaved throughout the site where irrigation is required. Drip irrigation systems can, therefore, also be moved and are not required, which is nice.

The design of drip irrigation provides maximum crop yield and increased fertilizer use on crops. Its localized supply of water results in reduced weed growth and also restricts the population of potential hosts. Drip irrigation systems result in minimal, if any, soil erosion as there is no surface runoff. This also controls potential fertilizer pollution in natural ground waters and surface waters. The use of emitters, control valves etc. allow the user to provide ready adjustment and sophisticated control of water supply to areas of the site. Seed germination is greatly improved and tillage operations are decreased.

Disadvantages of Drip Irrigation

There are many advantages for using drip irrigation over other irrigation methods and they are usually a great solution for commercial properties. Drip irrigation is not without its problems, as you'd expect. They do tend to require more maintenance than a more conventional system.

The slow water flow rate and low pressure can cause sediment to build up in pipes. Algae can even grow where climates permit. Mitigation of these issues requires regular flush outs of the system.

This is usually required at least annually but can be more frequent in the case of algal build up. Non-potable water contains more particles which can easily clog filters and drip emitters in particular. Drip emitter nozzles also require regular cleaning. These irrigation systems can also have an issue with salinity hazards.

Drip irrigation is best used for beds rather than say a lawn. Large open spaces that do require regular watering are better served with more conventional irrigation systems. For larger commercial applications, regular monitoring of plant health should be carried out to make sure the system is working at peak efficiency. Clogged or block emitters can cut off the water supply to "spots" in the field that will cause the gradual decline in health of the plants in the affected areas. This obviously adds an extra manpower cost to the facility. A well organized and managed monitoring system will identify problems early on, allowing for repairs to be carried out in a timely manner.

Water distribution elements of the system can also be damaged by exposure to sunlight especially if made of PVC. This can cause ongoing maintenance and repair costs that may not be the case with alternative irrigation systems.

Irrigation Efficiency

The ratio of the amount of water available (output) to the amount of water supplied (input) is known as Irrigation Efficiency. It is expressed in percentage.

Types of Irrigation Efficiency

The following are the various types of irrigation efficiencies:

Water Conveyance Efficiency (ηc)

It is the ratio of the amount of water applied, to the land to the amount of water supplied from the reservoir. It is obtained by the expression:

$$\eta c = Wl\ Wr \times 100$$

Where,

ηc =Water conveyance efficiency;

Wl = Amount of water applied to land;

Wr = Amount of water supplied from reservoir.

Water Application Efficiency (ηa)

It is the ratio of the water stored in root zone of plants to the water applied to the land. It is obtained by the expression:

$$\eta a = Wz\ Wl \times 100$$

Where,

ηa = Water application efficiency;

Wz = Amount of water stored in root zone;

Wl = Amount of water applied to land.

Water use Efficiency (ηu)

It is the ratio of the amount of water used to the amount of water applied. It is obtained by the expression:

$$\eta u = Wu\ Wl \times 100$$

Where,

ηu = Water use efficiency;

Wu = Amount of water used;

Wl = Amount of water applied to land.

Consumptive use Efficiency (ηcu)

It is the ration of the consumptive use of water to the amount of water depleted from the root zone. It is obtained by the expression:

$$\eta cu = Cu\ Wp \times 100$$

Where,

ηcu = Consumptive use efficiency;

Cu = Consumptive use of water;

Wp = Amount of water depleted from root zone.

Energy Efficient Irrigation

Efficient irrigation systems use energy-efficient equipment and designs, and also minimize the amount of unnecessary water use, adding to the energy savings. As a result, farms that irrigate efficiently will not only reduce their operating costs but will also reduce the use of water resources that are increasingly scarce. There are two main ways a farm can improve the efficiency of its irrigation efforts:

1. Improving the irrigation system, and
2. Enhancing the management and operations of the system.

Irrigation System Improvements

Drip irrigation delivers water only where it's needed, reducing water use and the energy required for pumping

Modifying irrigation systems can reduce energy and costs. For example, according to the Natural Resources Conservation Service (NRCS), in certain areas of the United States, switching from high- to low-pressure sprinkler systems can save as much as $55 and 770 kWh per acre annually. In areas where ground and surface water availability is diminishing, efficient irrigation tools such as drip trickle and lower-flow sprinkler systems save energy as well as water and money.

Some common causes of wasted energy in irrigation systems are worn or improperly sized pumps, worn nozzles, and improperly sized or designed fittings. Irrigation equipment problems and maintenance problems tend to go hand in hand. Pumps, motors, and engines that are badly designed or poorly maintained reduce the irrigator's degree of control over water applications, making it impossible to maintain correct soil moisture levels. This leads to crop stress, reduced yields, runoff, erosion, and other problems.

Irrigation Management Improvements

On the other hand, mechanical improvements alone do not necessarily bring energy savings. Better system performance typically causes higher pressure and increased volumes of applied water. These improvements should make it possible to meet crop water needs with fewer hours of irrigation. But if the irrigator continues to run the system for the same number of hours, energy consumption often stays the same or even increases.

In order to avoid both overwatering and underwatering, all irrigators need to know their system's net water application rate, measured in inches per hour or inches per irrigation. All irrigators should know general irrigation guidelines for the crops they grow. They should also know how to check their soil moisture levels. Most irrigators should also track crop water use, or evapotranspiration, as the season goes by.

Improving Water-efficient Irrigation

Agricultural irrigation is an energy intensive operation. Pressurized irrigation systems, especially center pivot sprinkler installations, use a high flow rate pump and require a large electric motor or engine. The major causes of increased energy use are associated with pipeline leaks, engine and pump efficiency and well maintenance. Poor uniformity of water application can also affect energy use by increasing pumping time. On center pivot systems, the major causes of poor water application uniformity are sprinkler nozzles that are worn or sized wrong, missing sprinkler heads, and leaking boots. Using a consistent method of irrigation scheduling during the growing season can optimize water application.

Improve Irrigation Efficiency

1. Use of a consistent method of irrigation scheduling can often reduce energy use by 7 to 30%. Using an ET-based irrigation scheduling system can ensure you are not under or overwatering the crop.
2. The average life expectancy of a sprinkler head is about seven to 10 years. The diameter of the sprinkler head nozzle is very important for uniform water application; and the nozzle diameter can grow with use, especially if there is sand or grit in the water. Poor application uniformity increases water pumping time and therefore energy use. Replace broken sprinkler heads as soon as possible. Do a "can test" to check the uniformity of the application pattern. Repair all leaks on the center pivot as soon as you notice them.
3. Buried pipelines rarely leak, unless they were not pumped out before winter. However, above ground pipelines frequently have worn gaskets and up to 30% of the water can be lost before it gets to the discharge point. Replace leaking gaskets and plug any holes in the pipeline.
4. The drawdown in a well increases if the screen becomes plugged. Increased drawdown greatly increases pumping costs. Screens become plugged due to mineral incrustation or from iron bacteria. Mineral incrustation occurs over time. By measuring the static and pumping water levels each year, the increase in drawdown can be measured and corrective action taken. Iron in the water usually means iron bacteria are present in the well. Annual chlorination will control the iron bacteria.
5. Maintain pumps regularly, including proper greasing and filling oil reservoirs every year. Adjust packing glands and adjusting impellers on deep well turbines regularly for efficient pump operation. Replace diesel engines with electric motors – that can have significant cost savings, depending on the price difference.
6. Most electric suppliers offer controlled (off-peak) electric rates for irrigation pumping systems. Using off-peak power rates can reduce pumping costs significantly when compared to regular power rates. However, off-peak rates should not be used with high-value crops like potatoes and onions. Talk with your electric supplier to determine if off-peak power rates would work for your operation. Typically, off-peak use will require a well capacity of 1400 gpm on a 130-acre center pivot or the capacity to irrigate in 100 hours per week. It works best for deep-rooted crops like corn or soybeans.

Soil Water Sensors

Soil water sensors have been used for irrigation and water management in agriculture for many years, but with limited success in many cases. Nonetheless, the use of soil water sensors is increasing as water scarcity increases and, conversely, as problems associated with over irrigation increase. Common problems with soil water sensing include sensor failure, problems with wiring, lack of or failure of data telemetry, inaccurate data, lack of timely data, excessive labor requirements and interference from dynamic soil temperature and bulk electrical conductivity changes. There are many sensors available, but only four main technologies: neutron thermalization, resistance blocks, capacitance sensing (frequency domain sensing), and travel time sensing (time domain reflectometry and time domain transmission modes). Understanding the theory of these offers insight into what a user can expect from each technology in terms of accuracy, stability and representativeness of the readings. The presentation will cover the types of sensors available, the operational theory of each sensor type, and explanations, with examples, of how the physical theory of operation dictates the limits of sensor calibration and performance, and of sensor representativeness in given soils.

Sprinkler Irrigation

In the sprinkler method of irrigation, water is sprayed into the air and allowed to fall on the ground surface somewhat resembling rainfall. The spray is developed by the flow of water under pressure through small orifices or nozzles. The pressure is usually obtained by pumping. With careful selection of nozzle sizes, operating pressure and sprinkler spacing the amount of irrigation water required to refill the crop root zone can be applied nearly uniform at the rate to suit the infiltration rate of soil.

Advantages of Sprinkler Irrigation

- Elimination of the channels for conveyance, therefore no conveyance loss.
- Suitable to all types of soil except heavy clay.
- Suitable for irrigating crops where the plant population per unit area is very high. It is most suitable for oil seeds and other cereal and vegetable crops.
- Water saving.
- Closer control of water application convenient for giving light and frequent irrigation and higher water application efficiency.
- Increase in yield.
- Mobility of system.
- May also be used for undulating area.
- Saves land as no bunds etc. are required.
- Influences greater conducive micro-climate.
- Areas located at a higher elevation than the source can be irrigated.
- Possibility of using soluble fertilizers and chemicals.
- Less problem of clogging of sprinkler nozzles due to sediment laden water.

Crop Response to Sprinkler

The trials conducted in different parts of the country revealed water saving due to sprinkler system varies from 16 to 70 % over the traditional method with yield increase from 3 to 57 % in different crops and agro climatic conditions.

Table: Response of different crops to sprinkler irrigation.

Crops	Water Saving, %	Yield increase, %
Bajra	56	19
Barley	56	16
Bhindi	28	23
Cabbage	40	3
Cauliflower	35	12
Chillies	33	24
Cotton	36	50
Cowpea	19	3
Fenugreek	29	35
Garlic	28	6
Gram	69	57
Groundnut	20	40
Jowar	55	34
Lucerne	16	27
Maize	41	36
Onion	33	23
Potato	46	4
Sunflower	33	20
Wheat	35	24

General classification of different types of sprinkler systems- Sprinkler systems are classified into the following two major types on the basis of the arrangement for spraying irrigation water:

1. Rotating head or revolving sprinkler system.
2. Perforated pipe system.

Rotating Head

Small size nozzles are placed on riser pipes fixed at uniform intervals along the length of the lateral pipe and the lateral pipes are usually laid on the ground surface. They may also be mounted on posts above the crop height and rotated through 90°, to irrigate a rectangular strip. In rotating type sprinklers, the most common device to rotate the sprinkler heads is with a small hammer activated by the thrust of water striking against a vane connected to it.

Few rotating type sprinkler irrigation systems

Perforated Pipe System

This method consists of drilled holes or nozzles along their length through which water is sprayed under pressure. This system is usually designed for relatively low pressure (1 kg/cm^2). The application rate ranges from 1.25 to 5 cm per hour for various pressure and spacing.

Based on the portability, sprinkler systems are classified into the following types:

1. Portable system: A portable system has portable main lines, laterals and pumping plant.

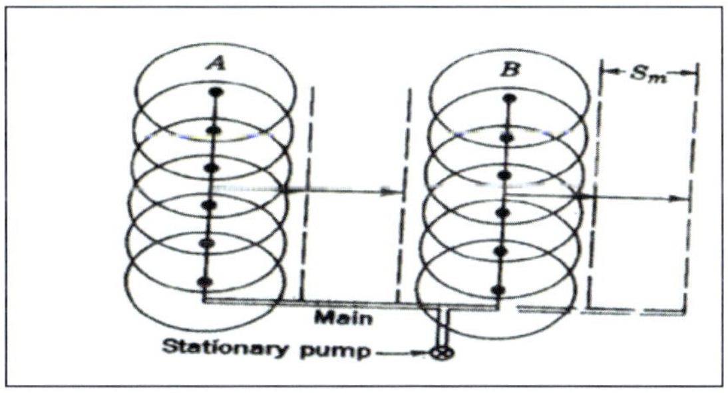

Fully portable sprinkler irrigation system

2. Semi portable system: A semi portable system is similar to a portable system except that the location of water source and pumping plant is fixed.
3. Semi-permanent system: A semi-permanent system has portable lateral lines, permanent main lines and sub mains and a stationery water source and pumping plant.
4. Solid set system: A solid set system has enough laterals to eliminate their movement. The laterals are positions in the field early in the crop season and remain for the season.
5. Permanent system: A fully permanent system consists of permanently laid mains, sub mains and laterals and a stationery water source and pumping plant.

Components of Sprinkler Irrigation System

The components of portable sprinkler system are shown through figure. A sprinkler system usually consists of the following components:

1. A pump unit,
2. Tubings- main/submains and laterals,

3. Couplers,
4. Sprinker head,
5. Other accessories such as valves, bends, plugs and risers.

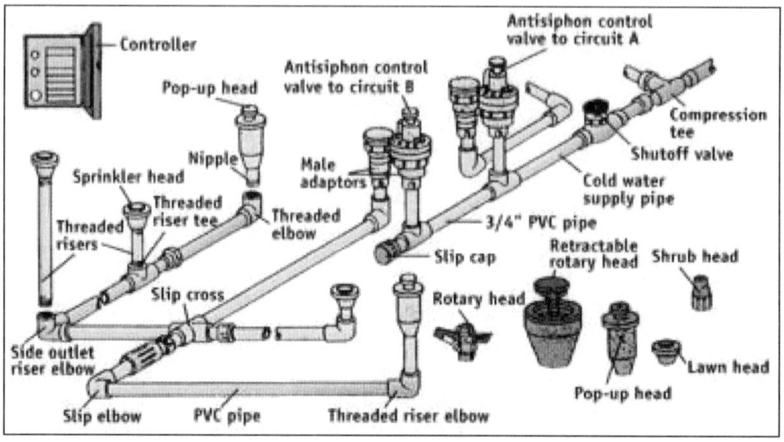

Component of a portable sprinkler irrigation system

1. Pumping Unit: Sprinkler irrigation systems distribute water by spraying it over the fields. The water is pumped under pressurc to thc fields. The pressure forces the water through sprinklers or through perforations or nozzles in pipelines and then forms a spray. A high speed centrifugal or turbine pump can be used for operating sprinkler irrigation for individual fields. Centrifugal pump is used when the distance from the pump inlet to the water surface is less than eight meters. For pumping water from deep wells or more than eight meters, a turbine pump is suggested. The driving unit may be either an electric motor or an internal combustion engine.

2. Tubings: Mains/submains and laterals: The tubings consist of mainline, submanins and laterals. Main line conveys water from the source and distributes it to the submains. The submains convey water to the laterals which in turn supply water to the sprinklers. Aluminum or PVC pipes are generally used for portable systems, while steel pipes are usually used for center-pivot laterals. Asbestos, cement, PVC and wrapped steel are usually used for buried laterals and main lines.

3. Couplers: Couplers are used for connecting two pipes and uncoupling quickly and easily. Essentially a coupler should provide:

a. A reuse and flexible connection,

b. Not leak at the joint,

c. Be simple and easy to couple and uncouple,

d. Be light, non-corrosive, durable.

4. Sprinkler Head: Sprinkler head distribute water uniformly over the field without runoff or excessive loss due to deep percolation. Different types of sprinklers are available. They are either rotating or fixed type. The rotating type can be adapted for a wide range of application rates and spacing. They are effective with pressure of about 10 to 70 m head at the sprinkler. Pressures ranging from 16 to 40 m head are considered the most practical for most farmers.

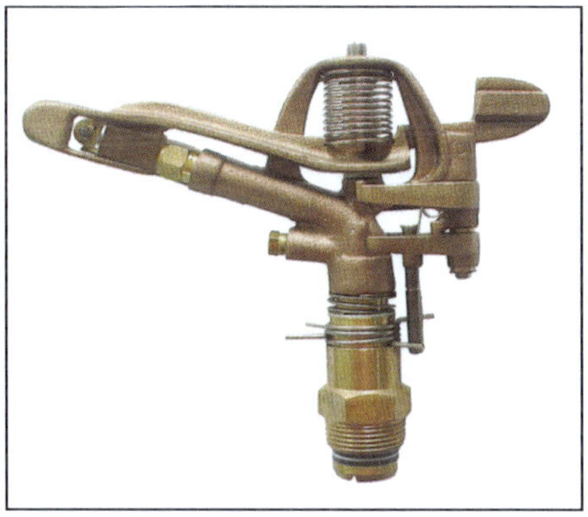

Sprinkler head

Fixed head sprinklers are commonly used to irrigate small lawns and gardens. Perforated lateral lines are sometimes used as sprinklers. They require less pressure than rotating sprinklers. They release more water per unit area than rotating sprinklers. Hence fixed head sprinklers are adaptable for soils with high intake rate.

5. Fittings and accessories: The following are some of the important fittings and accessories used in sprinkler system:

a. Water meters: It is used to measure the volume of water delivered. This is neccssary to operate the system to give the required quantity of water.

b. Flange, couplings and nipple used for proper connection to the pump, suction and delivery.

c. Pressure gauge: It is necessary to know whether the sprinkler system is working with desired pressure to ensure application uniformity.

d. Bend, tees, reducers, elbows, hydrants, butterfly valve and plugs.

e. Fertilizer applicator: Soluble chemical fertilizers can be injected into the sprinkler system and applied to the crop. The equipment for fertiliser application is relatively cheap and simple and can be fabricated locally. The fertilizer applicator consists of a sealed fertilizer tank with necessary tubings and connections. A venturi injector can be arranged in the main line, which creates the differential pressure suction and allows the fertilizer solution to flow in the main water line.

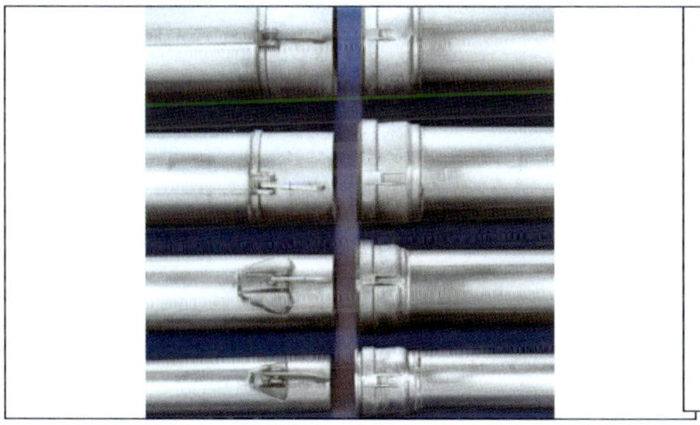

QRC HDPE sprinkler with metal latches

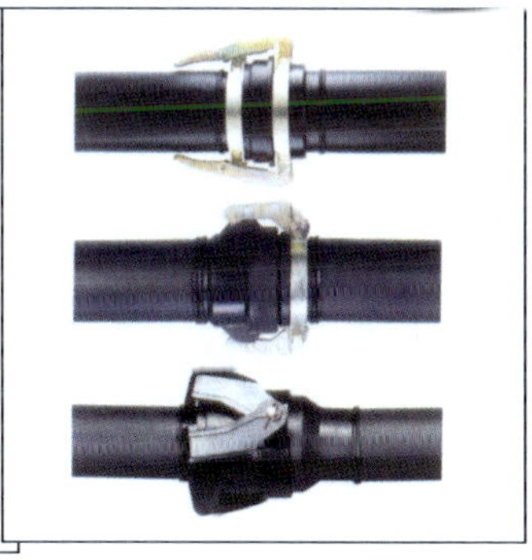

QRC HDPE sprinkler with plastic latches

QRC HDPE sprinkler base

QRC HDPE pump connecto

QRC HDPE bend

QRC HDPE tee

QRC HDPE end plug

Different sprinkler pipes and fittings

General Rules for Sprinkler System Design

- Main should be laid up and down hill:
  - Lateral should be laid across the slope or nearly on the contour.
  - For multiple lateral operations, lateral pipe sizes should not be more than two diameters.
  - Water supply source should be nearest to the center of the area.
- Layout should facilitate and minimize lateral movement during the season:
  - Booster pump should be considered where small portion of field would require high pressure at the pump.
  - Layout should be modified to apply different rates and amounts of water where soils are greatly different in the design area.

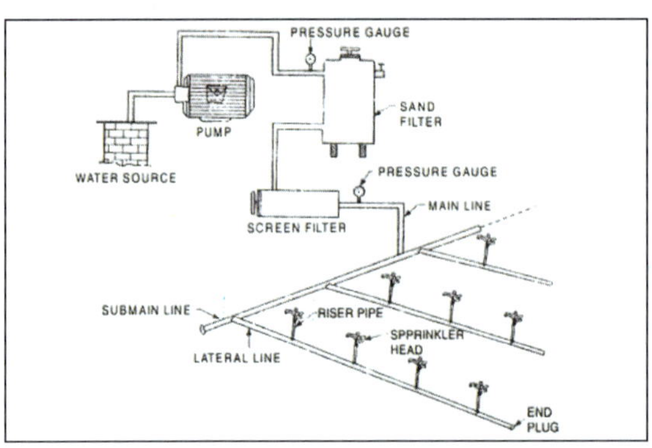

Layout of sprinkler irrigation system

Selecting the most Appropriate Sprinkler Systems

While selecting a sprinkler system, the most important physical parameters to be considered are:

1. The crop or crops to be cultivated.
2. The shape and size (acres) of the field.
3. The topography of the field.
4. The amount of time and labor required to operate the system.

Selecting Sprinkler System Capacity

A sprinkler system must be designed to apply water uniformly without runoff or erosion. The application rate of the sprinkler system must be matched to the infiltration rate of the most restrictive soil in the field. If the application rate exceeds the soil intake rate, the water will runoff the field or relocate within the field resulting in over and under watered areas.

The sprinkler system capacity is the flow rate needed to adequately irrigate an area and is expressed in liters per minute per acre. The system capacity depends upon on the; Peak crop water requirements during the growing season; effective crop rooting depth; texture and infiltration rate of the soil; the available water holding capacity of the soil; pumping capacity of the well or wells (if wells are the water source).

Constraints in Application of Sprinkler Irrigation

1. Uneven water distribution due to high winds,
2. Evaporation loss when operating under high temperatures,
3. Highly impermeable soils are not suitable,
4. Initial cost is high,
5. Proper design,
6. Lack of Package of practices,
7. Lack of awareness,
8. Lack of social concern to save natural resources,
9. High water pressure required in sprinkler ($>2.5 kg/cm^2$),
10. Difficulty in irrigation during wind in sprinkler.

Operation and Maintenance of Sprinkler Systems

Proper design of a sprinkler system does not in itself ensure success. It should be ensured that the prime mover and the pump are in alignment, particularly in the case of tractor-driven pumps. For these the drive shaft as well as the pump shaft should lie at nearly the same height to prevent too great an angle on the universal shaft.

While laying the main and lateral pipes, always begin laying at the pump. This necessarily gives the correct connection of all quick coupling pipes. While joining couplings, it is ensured that both the couplings and the rubber seal rings are clean.

In starting the sprinkler system, the motor or engine is started with the valves closed. The pump must attain the pressure stated on type-plate or otherwise there is a fault in the suction line. After the pump reaches the regulation pressure, the delivery valve is opened slowly. Similarly, the delivery valve is closed after stopping the power unit.

The pipes and sprinkler-lines are shifted as required after stopping.

Maintenance

General principles regarding the maintenance of the pipes and fittings and sprinkler heads are given below:

Pipes and Fittings

The pipes and fittings require virtually no maintenance but attention must be given to the following procedures:

a. Occasionally clean any dirt or sand out of the groove in the coupler in which the rubber. sealing ring fits. Any accumulation of dirt or sand will affect the performance of the rubber sealing ring.

b. Keep all nuts and bolts tight.

c. Do not lay pipes on new damp concrete or on piles of fertilizer. Do not lay fertilizer sacks on the pipe.

Sprinkler Heads

The sprinkler heads should be given the following attention:

a. When moving the sprinkler lines, make sure that the sprinklers are not damaged or pushed into the soil.

b. Do not apply oil, grease or any lubricant to the sprinklers. They are water lubricated and using oil, grease or any other lubricant may stop them from working.

c. Sprinklers usually have a sealed bearing and at the bottom of the bearing there are washers. Usually it is the washers that wear and not the more expensive metal parts. Check the washers for wear once a season or every six months which is especially important where water is sandy. Replace the washers if worn.

d. After several season's operation the swing arm spring may need tightening. This is done by pulling out the spring end at the top and rebending it. This will increase the spring tension.

In general, check all equipment at the end of the season and make any repairs and adjustments and order the spare parts immediately so that the equipment is in perfect condition to start in the next season.

Storage

The following points are to be observed while storing the sprinkler equipment during the off season:

a. Remove the sprinklers and store in a cool, dry place.

b. Remove the rubber sealing rings from the couplers and fittings and store them in a cool, dark place.

c. The pipes can be stored outdoors in which case they should be placed in racks with one end higher than the other. Do not store pipes along with fertilizer.

d. Disconnect the suction and delivery pipe-work from the pump and pour in a small quantity of medium grade oil. Rotate the pump for a few minutes. Blank the suction and delivery branches. This will prevent the pump from rusting. Grease the shaft.

e. Protect the electric motor from the ingress of dust, dampness and rodents.

Trouble Shooting

The following are the general guidelines to identify and remove the common troubles in the sprinkler systems:

1. Pump does not prime or develop pressure:

a. Check that the suction lift is within the limits. If not, get the pump closer to the water.

b. Check the suction pipeline and all connections for air leaks. All connections and flanges should be air tight.

c. Check that the strainer on the foot valve is not blocked.

d. Check that the flap in the foot valve is free to open fully.

e. Check the pump gland (s) for air leaks. If air leaks are suspected tighten the gland (s) gently. If necessary repack the gland (s) using a thick grease to seal the gland satisfactorily.

f. Check that the gate valve on the delivery pipe is fully closed during priming and opens fully when the pump is running.

2. Sprinklers do not turn:

a. Check pressure.

b. Check that the nozzle is not blocked. Preferably unscrew the nozzle or use a small soft piece of wood to clear the blockage. Do not use a piece of wire or metal as this may damage the nozzle.

c. Check the condition of washers at the bottom of the bearing and replace them if worn or damaged.

d. Check that the swing arm moves freely and that the spoon which moves into the water stream is not bent by comparing it with a sprinkler which is operating correctly.

e. Adjust the swing arm spring tension: Usually it should not be necessary to pull up the spring by more than about 6 mm.

3. Leakage from coupler or fittings: The sealing rings in the couplers and fittings are usually designed to drain the water from the pipes when the pressure is turned off. This ensures that the pipes are automatically emptied and ready to be moved. With full pressure in the system the couplers and fittings will be effectively leak-free. If, however, there is a leakage, check the following:

a. There is no accumulation of dirt or sand in the groove in the coupler in which the sealing ring fits. Clean out any dirt or sand and refit the sealing ring.

b. The end of the pipe going inside the coupler is smooth, clean and not distorted.

c. In the case of fittings such as bends, tees and reducers ensure that the fitting has been properly connected into the coupler.

Digital Irrigation Flow Meter Sensor

In the past, mechanical type water flow meters are widely used for irrigation, but mechanical flow meters have many inherent defects. As a result, more and more digital water flow meters are used for irrigation application.

Digital irrigation flow meters mainly have transit-time ultrasonic flow meters and magnetic flow meters now which have no moving parts. Of course, it doesn't means only the 2 type available in the market, infact, there are also other type digital water flow meters available, but ultrasonic and magnetic meters almost most of markets with their economic price and good performance.

Also because of human resource, intelligent automatically controlling is also listed as a priority consideration. Digital water flow meter, e.g. ultrasonic type, can provide versatile outputs for remote signal controller, and have 4-20mA, OCT pulse, relay switch, batch controller, etc. By different signal output, users can reach different intelligent controlling as requirement.

Digital water flow meters have been widely used for agriculture and turf, and typical application including the following:

a. Agriculture fertigation

b. Landscaping

c. Golf Watering

d. Pivot irrigation

e. Irrigation sprinkler

The following are 8 kinds of main solutions:

1. Solar Powered Clamp on Ultrasonic Flow Meter: Clamp on ultrasonic flow meter doesn't require to cut pipe and easy to install & operate. But clamp on flow meter may cost power much sharply

than inline and insertion type. So built-in battery is not good solution for several days flow measurement. But if you just require a day measurement, our portable flow meter with battery is a good choice. Other a storage cell is required. If just want the flow meter to work for several days, e.g. a week, a storage cell should be enough. But if want to work for more days or several months, solar power solution should be a better choice.

2. Battery Power Inline Ultrasonic Water Meter: With cutting-edge technology, we have developed our transit-time ultrasonic water flow meter with very low power consumption. Thus this patented technology makes our inline ultrasonic flow meters powered by built-in 3.6V Li-battery. Such battery can support ultrasonic water meter to work for 4-6 years, even longer, so the solution doesn't require external storage cell or solar panel. Battery powered ultrasonic water meters have been widely used in irrigation projects.

3. Battery Powered Ultra-light Ultrasonic Water Meter: In order to reduce the weight of irrigation flow meter, we specially designed ultra-light insertion ultrasonic water flow meters for irrigation. It has a compact body and very light weight. It is also powered by built-in 3.6V battery and can work for 4-6 years. The ultrasonic water meter has four pipe sizes released: DN80, DN100, DN125 and DN150mm. Comparing with inline ultrasonic water meter, its price is much cheaper. Thus, it is an ideal type for your irrigation flow measurement.

4. Ultrasonic Portable Water Flow Meter: Portable water flow meters are also used for irrigation flow measurement, especially when the irrigation process is for short-term and often moved for

different position. Portable digital water flow meter is developed on the basis of clamp on transit-time ultrasonic flow meter. It has built-in recharging battery and can work 12-20 hours after fully recharging. Meanwhile, clamp on flow transducers don't need to cut pipe. The whole weight of portable flow meter is not big and easy to carry for mobile flow check and survey.

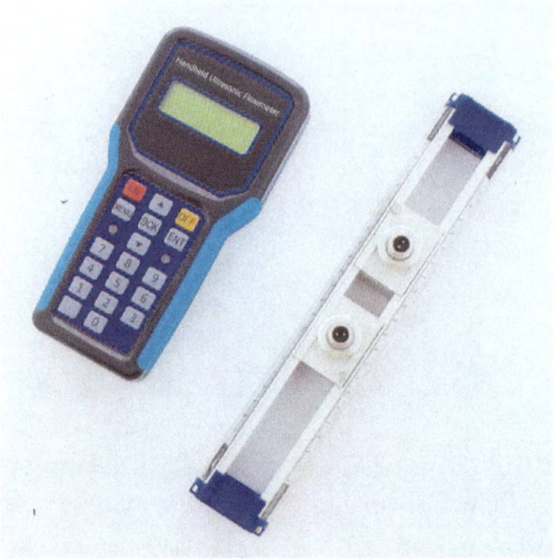

5. Battery Power Inline Magnetic Flow Meter: As the main competitor of ultrasonic water meter, battery power magnetic flow meters are also often selected as digital irrigation flow meters. But comparing with battery powered ultrasonic water meter, its price is much higher. Meanwhile, because of different structure, the weight of mag-flow is bigger than ultrasonic type. This will improve shipping fee, especially for international business. Thus, out of comprehensive consideration, it is firstly to recommend ultrasonic water meters.We can make the two types flow meters and provide them as your requirement. But we recommend ultrasonic type as first choice.

6. Plastic Inline Turbine Flow Meter: This solution has moving parts and not recommended. But there are still many suppliers to provide such solution. End users are not familiar with flow technologies, but as distributors or manufacturers, the solution shouldn't be recommended unless requirement.

Anyway, in order for such requirement, we also provide plastic type turbine flow meters. Plastic body can reduce the weight. The turbine flow meter uses impeller rotation to 'sense" water flow, thus its working doesn't require external power. But if require signal output, its converter will require, but in regard that it is just signal, a battery power should be enough. If battery power costs up, user can replace a new one.

7. Paddle wheel Irrigation Flow Sensor with Pulse Output: Our paddle wheel flow sensor is not heavy and has pulse output. Meanwhile, its price is not higher. Thus, paddle flow meter sensor is often selected as irrigation flow sensor, especially for bigger pipe size. We also have developed related digital flow controller to make the flow sensor as digital irrigation flow meter.

8. Inline Hall Effect Flow Sensor: Many irrigation projects are limited by budget target and have no enough budgets to select irrigation flow meters above. Thus, we have released inline hall effect flow sensor with pulse output. Comparing with flow devices above, its main advantage is cheap. In order to meet requirement for bigger pipe size, we have made DN80mm inline hall flow sensor which is a bigger size in such flow sensors.

Of course, comparing with the flow meters above, its accuracy is not so high like them, but the accuracy should be enough for common irrigation application.

Landscaping and Turf Irrigation Flow Meter Sensor

Unlike agriculture irrigation, turf and landscaping irrigation may use smaller size pipe. e.g. below

DN40mm. For such application, we can provide small size digital water flow meter or water flow sensor, which mainly includes:

- Hall effect flow sensors.
- Plastic variable area flow meter.
- Turbine flow sensors.
- Ultrasonic flow sensor.
- Other customizing flow devices.

In a word, ultrasonic flow meter and magnetic flow meter should be the main irrigation flow meters at present. When you need to purchase flow meters for irrigation watering, it is better to select either as your first choice.

Importance of Digital Water Flow Meter for Irrigation

Water has been one of the most precious resources in the world, and more and more countries and areas are facing serious problem of lacking water. Thus, a precise irrigation flow meter is essential to measure how much water rate you have consumed. Meanwhile, as agricultural irrigation has its unique application: irrigation flow meters may be carried for different of applications in different farm, so its weight shouldn't be too heavy.

In the past, mechanical water meters or flow meters were used for irrigation flow measurement and mainly include woltman water meters, turbine flow meters, etc. Such traditional flow meters always have moving parts and brought many disadvantages: pressure loss, shorter life, accuracy reduced after using, bigger weight, etc. Thus, it is absolute to introduce proper irrigation flow meters. So digital water flow meters are released to avoid those problems.

Transit-time ultrasonic flow meters and magnetic flow meters are the better solution for digital irrigation flow metering. Comparing with mechanical irrigation flow meters, they have very good advantages as below:

- Higher accuracy,
- Lower velocity sensitivity,
- Longer using life obviously reduces your average investigation,
- Powerful function to record positive/negative/NET flow rate,
- Stronger communication ability: batch controller, data logger, GPRS/GSM wireless communication, Modbus for network, etc.,
- Clamp on ultrasonic flow meters don't need to cut pipe and lightweight to carry for different test points. Reduce your comprehensive investigation,
- Low cost clamp on flow meter solution with cutting edge technology, we have reduced the price of clamp on flow meter to a very low point that will give you a big surprise.

Variable Rate Spraying

Variable rate spray application is receiving a lot or attention with our increased ability to farm according to prescription maps. For dry products such as seed or fertilizer, metering is relatively straight-forward and variable rate application has been possible for many years. However, liquid product application has been more complex and requires special approaches.

Hydraulic Pressure

In conventional liquid metering, the liquid is forced through a metering orifice that is placed in-line. This could be an orifice plate for liquid fertilizer, or a flat fan nozzle for pesticides. Rate control is achieved by altering the spray pressure. It is usually impractical to change the nozzle or metering orifice during an application.

The main drawback to this approach is that spray pressure is not very effective at changing flow rates due to the square root relationship between spray pressure and flow rate.

For example, with reference to the table below, one can see that doubling the spray pressure (say, from 30 to 60 psi) only increases the flow rate by 40%. Tripling the pressure (from 30 to 90 psi) increases the application volume by 73% (we can call that a factor of 1.73). As a result, the use of pressure alone doesn't offer a large range of application rates, and we accept a factor of 2 to be the limit for fertilizer streamer and broadcast nozzles (meaning a four-fold pressure range) and a factor of 1.73 to be practical for broadcast pesticide sprays over a 3-fold pressure range. Any wider application volume range would require adjustment to travel speed.

Nozzle Size	Pressure (psi)	Nozzle Flow (US gpm)	Travel Speed (columns, mph) at listed application volume (headings, US gal/acre)											
			3	4	5	6	7	8	9	10	12	14	16	18
			US gpa											
	30	0.130	12.9	9.6	7.7	6.4	5.5	4.8	4.3	3.9	3.2	2.8	2.4	2.1
	40	0.150	14.9	11.1	8.9	7.4	6.4	5.6	5.0	4.5	3.7	3.2	2.8	2.5
015	50	0.168	16.6	12.5	10.0	8.3	7.1	6.2	5.5	5.0	4.2	3.6	3.1	2.8
	60	0.184	18.2	13.6	10.9	9.1	7.8	6.8	6.1	5.5	4.5	3.9	3.4	3.0
Green	70	0.198	19.6	14.7	11.8	9.8	8.4	7.4	6.5	5.9	4.9	4.2	3.7	3.3
	80	0.212	21.0	15.8	12.6	10.5	9.0	7.9	7.0	6.3	5.3	4.5	3.9	3.5
	90	0.225	22.3	16.7	13.4	11.1	9.5	8.4	7.4	6.7	5.6	4.8	4.2	3.7
	100	0.237	23.5	17.6	14.1	11.7	10.1	8.8	7.8	7.0	5.9	5.0	4.4	3.9
	30	0.173	17.1	12.9	10.3	8.6	7.3	6.4	5.7	5.1	4.3	3.7	3.2	2.9
	40	0.200	19.8	14.9	11.9	9.9	8.5	7.4	6.6	5.9	5.0	4.2	3.7	3.3
02	50	0.224	22.1	16.6	13.3	11.1	9.5	8.3	7.4	6.6	5.5	4.7	4.2	3.7
	60	0.245	24.2	18.2	14.5	12.1	10.4	9.1	8.1	7.3	6.1	5.2	4.5	4.0
Yellow	70	0.265	26.2	19.6	15.7	13.1	11.2	9.8	8.7	7.9	6.5	5.6	4.9	4.4
	80	0.283	28.0	21.0	16.8	14.0	12.0	10.5	9.3	8.4	7.0	6.0	5.3	4.7
	90	0.300	29.7	22.3	17.8	14.9	12.7	11.1	9.9	8.9	7.4	6.4	5.6	5.0
	100	0.316	31.3	23.5	18.8	15.7	13.4	11.7	10.4	9.4	7.8	6.7	5.9	5.2
	30	0.217	21.4	16.1	12.9	10.7	9.2	8.0	7.1	6.4	5.4	4.6	4.0	3.6
	40	0.250	24.8	18.6	14.9	12.4	10.6	9.3	8.0	7.4	6.2	5.3	4.6	4.1
025	50	0.280	27.7	20.8	16.6	13.8	11.9	10.4	9.2	8.3	6.9	6.0	5.2	4.6
	60	0.306	30.3	22.7	18.2	15.2	13.0	11.4	10.1	9.1	7.6	6.5	5.7	5.1
Lilac	70	0.331	32.7	24.6	19.6	16.4	14.0	12.3	10.9	9.8	8.2	7.0	6.1	5.5
	80	0.354	35.0	26.3	21.0	17.5	15.0	13.1	11.7	10.5	8.8	7.5	6.6	5.8
	90	0.375	37.1	27.8	22.3	18.6	15.9	13.9	12.4	11.1	9.3	8.0	7.0	6.2
	100	0.395	39.1	29.3	23.5	19.6	16.8	14.7	13.0	11.7	9.8	8.4	7.3	6.5

With these inherent limitations in flow rate capacities from hydraulic pressure alone, applicators are often forced to use wide pressure fluctuations to achieve reasonable rate responses. In some cases, this means that pressure needs can be too low for uniform distribution, or too high for pump or plumbing capacitles.

There are a few options available that expand the flow rate range of liquid products. Some of these have been available for a number of years; others are just entering the market for 2016. A brief overview of the main options follows:

VariTarget Nozzle

This nozzle design uses a spring-loaded plunger to exert force on a flexible nozzle cap, deflecting it slightly. The deflection changes the orifice size, allowing for a change in flow. As a result, the flow rate response to a pressure change is increased dramatically. A single VariTarget nozzle equipped with a blue or green nozzle cap can deliver flows ranging from 0.2 US gpm at 20 psi to 1.2 gpm at 65 psi, for a stunning 6-fold change in application rate.

The main drawback of this nozzle is the poor metering accuracy of the system. In calibration tests, flows from various new VariTarget nozzles operated at the same pressures varied by more than 10% in many cases. While this amount of variability may be acceptable in liquid fertilizer application, it is not considered acceptable for pesticide application.

Application Chart Blue Tip

Reference Pressure PSI	*Nozzle Flow Rate GPM*	20" Nozzle Spacing							
		US Gallons / Acre - Based on Water							
		4 MPH	6 MPH	8 MPH	10 MPH	12 MPH	14 MPH	16 MPH	20 MPH
16	0.15	11.1	7.4	5.6	4.5	3.7	3.2	2.8	2.2
20	0.20	14.9	9.9	7.4	5.9	5.0	4.2	3.7	3.0
32	0.30	22.3	14.9	11.1	8.9	7.4	6.4	5.6	4.5
35	0.40	29.7	19.8	14.9	11.9	9.9	8.5	7.4	5.9
38	0.50	37.1	24.8	18.6	14.9	12.4	10.6	9.3	7.4
41	0.60	44.6	29.7	22.3	17.8	14.9	12.7	11.1	8.9
44	0.70	52.0	34.7	26.0	20.8	17.3	14.9	13.0	10.4
47	0.80	59.4	39.6	29.7	23.8	19.8	17.0	14.9	11.9
60	1.00	74.3	49.5	37.1	29.7	24.8	21.2	18.6	14.9
65	1.20	89.1	59.4	44.6	35.6	29.7	25.5	22.3	17.8
85	1.50	111.4	74.3	55.7	44.6	37.1	31.8	27.8	22.3

TeeJet Variable Rate Fertilizer Assemblies

These metering assemblies, new for 2016, offer an elastomer (EPDM) metering plate whose orifice diameter expands with pressure, offering a wider range of flows.

There are no moving parts in the assembly. Four models are available:

- PTC-VR: Using a push-to-connect design for planters and toolbars, it offers versions that accomodate 1/4″, 5/16″, and 3/8″ OD tubing diameters.
- QJ-VR Hose Barb: This unit offers hose barb diameters for 1/4″ and 3/8″ ID hose.

Both units feature a pressure range of 10 psi to 100 psi, within which a flow rate range of approximately 8-fold is possible:

- SJ3-VR: This unit generates three streams and operates over a pressure range of 20 to 100 psi, offering a flow rate range of about 4.5.
- SJ7-VR: Generating seven streams and operating over a pressure range of 30 to 80 psi, this unit allows a flow rate range of about 2.9.

In all cases, the realized flow rate range is significantly greater than would have been achieved with pressure change alone. TeeJet has tested the flow rate variance among units operating at the same pressure and has found them to be acceptable, according to company representatives.

Hypro VFR ESI

This 6-stream fertilizer nozzle, set for release in 2016 or 2017, has an internal mechanism similar to the VariTarget nozzle design, using a spring-loaded mechanism to change the metering orifice size. This nozzle does not use an elastomer orifice, and its metering is proving to be accurate and repeatable. A three-fold flow rate change is achievable at pressures ranging from 15 to 60 psi. Three sizes will be available: VFR ESI 015-04, VFR ESI 5-08, and VFR ESI 10-20. Each will feature a pressure range of 15 ro 60 psi, within which it offers a 3-fold flow rate range.

Fertilizer banding has greater tolerances for application because pattern width is less important, and also because stream stability is less affected by pressure than spray pattern droplet size.

Pulse Width Modulation

PWM utilizes conventional plumbing: A single boom line and a single nozzle at each location. Liquid flow rate through each nozzle is managed via an intermittent, brief shutoff of the nozzle flow activated by an electric solenoid that replaces the spring-loaded check valve. Typical systems pulse at 10 Hz (the solenoid shuts off the nozzle 10 times per second), and the duration of the nozzle in the "on" position is called the duty cycle (DC) or pulse width.

100% DC means the nozzle is fully on, and 20% DC means the solenoid is open only 20% of the time, resulting in the nozzle flowing at approximately 20% of its capacity. This is illustrated in the figure below. The ability to control the duty cycle is referred to as pulse width modulation.

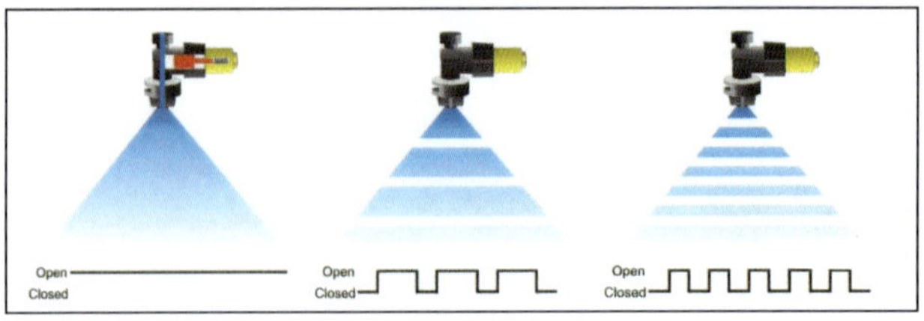

The system has a theoretical flow rate range of about four- to five-fold. Within this range, spray pressure, and the corresponding spray pattern and droplet size, stay roughly constant. This makes it ideal for variable rate pesticide application, where spray patterns and spray quality are critical for performance.

The main disadvantage of this system is cost. Although highly accurate and dependable, commercial sprayer units are priced between $15,000 and $30,000 per sprayer, depending on features and boom widths. Competition may lower costs, with three systems available for 2016.

For ammonia and liquid fertilizer seeders or toolbars, Capstan recommends different PWM products, N-Ject NH3 or N-Ject LF. These systems offer more control over PWM pulse frequency and duty cycle and can achieve over 10-fold rate ranges. At low frequencies and duty cycles, the mobiliy of the fertilizer in soil needs to be considered, as significant gaps in a stream can be generated.

Dual Boom Systems

A second boom fitted with different flow nozzles is installed, and is activated when the flow rate requirements can no longer be met with a single set of nozzles. Once the second boom is activated, the spray pressure drops significantly and additional speed capacity can be realized.

Dual or Quadruple Nozzle Bodies

A similar approach to the dual boom is available from Arag (Seletron) or Hypro (Duo React). These systems utilize a single boom and direct the flow through one of any two or four (Arag Seletron only) nozzles, or several nozzles at the same time. Similar pressure fluctuations as with a dual boom would be experienced, requiring careful selection of nozzle flow rates to avoid large pressure jumps. The system can also be used to manually change from one nozzle to another as needed. In the figure below, the pressure changes associated with the sequential use of 015, 02, and 035 flows are shown.

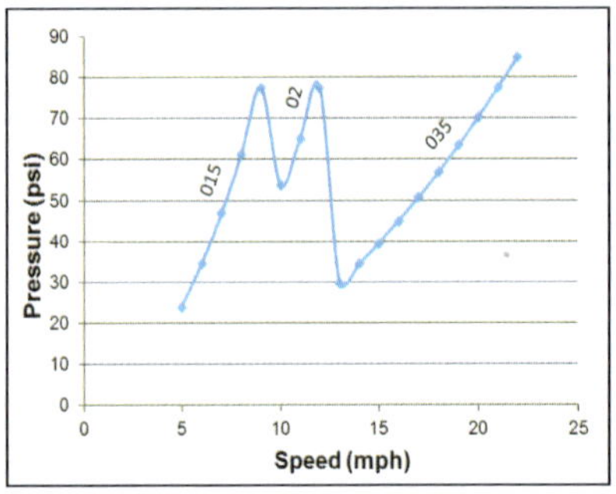

Twin Fluid Nozzles

A single nozzle uses both liquid and air pressure to control liquid flow rate. Liquid is delivered to the nozzle in the usual way, by pressure. Air is also delivered, created by a dedicated air pump that has modest volume and pressure requirements. Both air and liquid make their way through the same nozzle (a deflector style, similar to the TeeJet FloodJet).

The advantage is that liquid flow, and droplet size, can be adjusted with air and liquid pressure. More air results in lower liquid flow and reduces droplet size. More liquid pressure increases flow. Clever combinations of both can keep droplet size fairly constant over a wide flow rate range. Alternatively, the nozzles can change droplet size while keeping the same flow rate, depending on the drift or coverage needs at the time.

Direct Injection

Direct injection is an option for variable application of pesticides. In this system, undiluted pesticide is placed into canisters on the sprayer, and plain water (or water plus adjuvant) is in the sprayer tank. The chemical is metered and introduced into the water on the pressure side at some distance upstream from the boom sections. The pesticide rate can be varied with the speed of the direct injection pump, offering a very high dynamic range of possible rates. For example, Raven's Sidekick Pro offers a 40-fold range of flow rates.

After injection, an in-line mixer ensures that products are evenly distributed in the carrier. The amount of lag in the systems will depend on the amount of spray mixture in the plumbing up stream of the nozzles, the total boom flow rate, as well as the boom section configuration. With a variable rate map this lag can be anticipated and accommodated.

Pump technology has improved the metering accuracy over a range of viscosities. However, dry formulations remain a challenge as slurries can settle and create problems for the pump and screen components.

Automatic Irrigation

Automatic irrigation is the use of a device to operate irrigation structures so the change of flow of water from one bay, or set of bays, to another can occur in the absence of the irrigator.

Automation can be used in a number of ways:

- To start and stop irrigation through supply channel outlets,
- To start and stop pumps,
- To cut off the flow of water from one irrigation area – either a bay or a section of channel and directing the water to another area.

These changes occur automatically without any direct manual effort, but the irrigator may need to spend time preparing the system at the start of the irrigation and maintaining the components so it works properly.

Benefits of Automatic Irrigation

Reduced Labour

As the irrigator is not required to constantly monitor the progress of an irrigation, thc irrigator is available to perform other tasks – uninterrupted.

Improved Lifestyle

The irrigator is not required to constantly check the progress of water down the bays being irrigated. The irrigator is able to be away from the property, relax with the family and sleep through the night.

- More timely irrigation: Irrigators with automation are more inclined to irrigate when the plants need water, not when it suits the irrigator.
- Assists in the management of higher flow rates: Many irrigators are looking to increase the irrigation flow rates they receive through installing bigger channels and bay outlets. Such flow rates generally require an increase in labour as the time taken to irrigate a bay is reduced thus requiring more frequent change over. Automation allows for these higher flows to be managed without an increase in the amount of labour.
- More accurate cut-off: Automation of the irrigation system allows cut-off of water at the appropriate point in the bay. This is usually more accurate than manual checking because mistakes can occur if the operator is too late or too early in making a change of water flow.
- Reduced runoff of water and nutrients: Automation can help keep fertiliser on farm by effectively reducing run off from the property. Retaining fertiliser on farm has both economic and environmental benefits.
- Reduced costs for vehicles used for irrigation: As the irrigator is not required to constantly check progress of an irrigation, motor bikes, four wheelers and other vehicles are used less. This reduces the running costs of these vehicles and they require less frequent replacement.

Disadvantages of Automatic Irrigation

1. Cost: There are costs in purchasing, installing and maintaining automatic equipment.
2. Reliability: Can the irrigator trust an automatic system to work correctly every time? Sometimes failures will occur. Often these failures are because of human error in setting and maintaining the systems. A re-use system is good insurance to collect any excess runoff when failures occur.
3. Increased channel maintenance: There is a need to increase maintenance of channels and equipment to ensure the system works correctly. Channels should be fenced to protect the automatic units from stock damage.

Types of Automatic Irrigation Systems Available

1. Pneumatic system: A pneumatic system is a permanent system activated by a bay sensor located at the cut-off point. When water enters the sensor, it pressurizes the air, which is piped to a mechanism that activates the opening and closing of irrigation structures.
2. Portable timer system: A portable timer system is a temporary system which uses electronic clocks to activate the opening and closing of the irrigation structures. Because of its portable nature, landowners usually buy 4 or 5 units to move around the whole property.
3. Timer/ Sensor Hybrid: As the name suggests, this system is a hybrid of portable timer and sensor systems. Like a portable timer, it uses an electronic device to activate the opening and closing of the irrigation structures. However, this system has an additional feature of the irrigator being able to place a moveable sensor down the bay, which when comes in contact with water, transmits radio signals to the timer devices at the outlets to open or close the structures and sends a radio message to a receiver to let the landowner know water has reached the cut-off points down the bay.

SCADA

Automation systems that use Supervisory Control and Data Acquisition (SCADA) consist of a personal computer and software package to schedule and control irrigation via a radio link. Signals are sent from the computer to control modules in the paddock to open and close irrigation structures with linear actuators. Bays are opened and closed on a time basis, some systems have the capacity to automatically alter the time a bay outlet is open if the channel supply is inconsistent.

SCADA based systems have the additional benefit of being able to start and stop irrigation pumps and motors.

Automation Irrigation Layout

An irrigation layout can be automated at one of two places; in sections of channel or at individual bay outlets.

Automation of Channel Sections

In this system the channel structures are automated allowing the channel level to be changed. The bay outlets do not have opening or closing structures rather each set of outlets is set at a specific level eg a set of sills.

This method of automation requires a larger amount of fall to be available in the channel system to allow for a change in water level between different areas. This change in water level is required to prevent water flowing onto bays previously irrigated, when another section is to be irrigated. On many farms this fall is not available, so this method of automation in many cases is not suitable.

Automation of Individual Bay Outlets

This method of automation involves control of the bay outlets to change the flow of water onto the areas being irrigated. This system of automation is the most frequently used in areas where there is insufficient fall to automate channel sections.

The same type of automatic devices available can be set up to operate either automation of channel sections or automation of bay outlets.

Which System is Best

All systems of automation have advantages and disadvantages that need to be considered when deciding which system will suit the irrigation layout for a particular property. There is no system that will be the "best" system for all properties.

The methods of irrigation used by the irrigator need to be considered. If a system that can be moved around the property and perhaps used on other properties is required, then the irrigator needs to consider those systems that are portable. If the irrigator wants a system where the components are fixed and can follow the same irrigation sequence each irrigation, then a fixed system would be more appropriate.

In determining the best system for a property, the irrigator will need to consider the cost of the system, back up servicing of the system and which system will best suit the property and irrigation layout.

Automatic Plant Irrigation System Circuit and its Working

In present days, in the field of agriculture farmers are facing major problems in watering their crops. It's because they don't have proper idea about the availability of the power. Even if it is available, they need to pump water and wait until the field is properly watered, which compels them to stop doing other activities – which are also important for them, and thus they loss their precious time and efforts. But, there is a solution – an automatic plant irrigation system not only helps farmers but also others for watering their gardens as well.

This automatic irrigation system senses the moisture content of the soil and automatically switches the pump when the power is on. A proper usage of irrigation system is very important because the main reason is the shortage of land reserved water due to lack of rain, unplanned use of water as a result large amounts of water goes waste. For this reason, we use this automatic plant watering system, and this system is very useful in all climatic conditions.

Automatic Plant Irrigation System

Automatic Plant Irrigation System

The power supply consists of a step-down transformer, which steps down the voltage to 12VAC. By using a bridge rectifier this AC is converted to DC, then it is regulated to 5v using a voltage regulator which is used for the operation of the microcontroller.

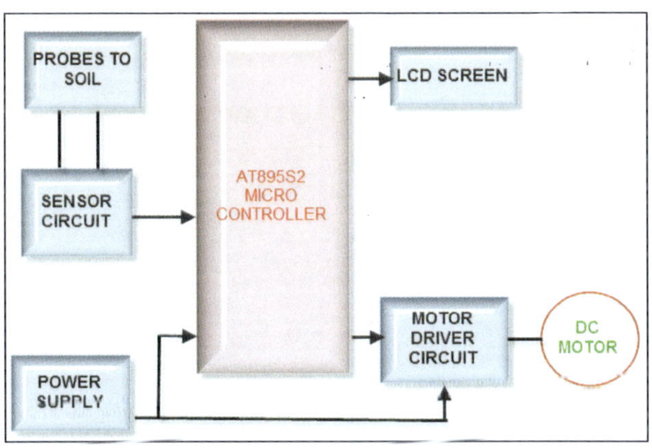

Block Diagram of Automatic Plant Irrigation System

The block diagram of this automatic plant irrigation system comprises three main components namely a microcontroller, a motor-driver circuit and a sensor circuit. When the sensor circuit senses the condition of soil, it compares it with the reference voltage 5v. This process is done by a 555 timer.

When the soil condition is less than the reference voltage, i.e. 5v, then the soil is considered as dry and instantly the 555 timer sends the logic signal 1 to the microcontroller. The microcontroller then turns on the motor driver circuit and prompts the motor to pump water to the plants. When the soil condition is greater than the reference voltage, the soil becomes dry. Then the timer sends the logic signal 0 to the microcontroller, this turns off the motor driver circuit and prompts motor to pump water to the fields. Finally, the condition of the motor and soil are displayed in the LCD display.

Circuit Diagram of Plant Irrigation System

The main component used in this automatic plant irrigation system is 7404 Hex Inverter. The main function of the inverter output is proportional to input. It means, if the input of the inverter is low, then the output of the inverter will be high, and the inverter will give low output if the input is high. The Hex inverter 7404 IC includes six independent inverters and the range of operating voltage is around 4.75V to 5.5V, and the Supply voltage is 5V. They are used in many applications such as drivers, inverting buffers, etc. This IC is available in different packages like quad-flat package and dual-inline package. The pin configuration of the 7404 IC is shown in figure below.

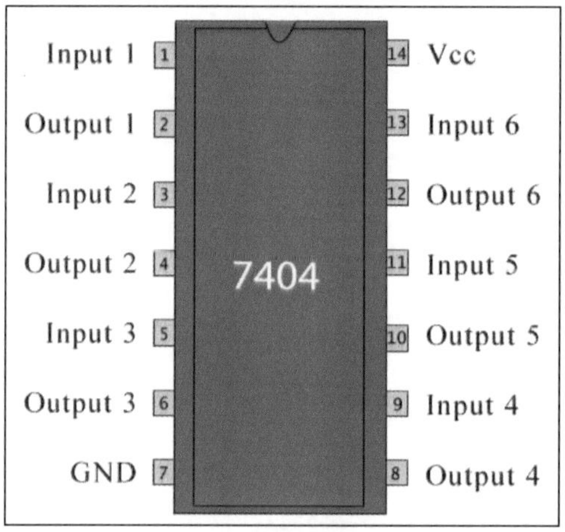

7404 IC Pin Configuration

The circuit diagram of the plant-irrigation system is shown below. To make the circuit work and to water the pants, we use this simple logic: when the soil is dry, it has high resistance and when the soil is wet it has low resistance. This circuit consists of two probes that are placed into the earth. These probes perform the work only when the soil resistance is low and they cannot perform when the resistance of the soil is high.

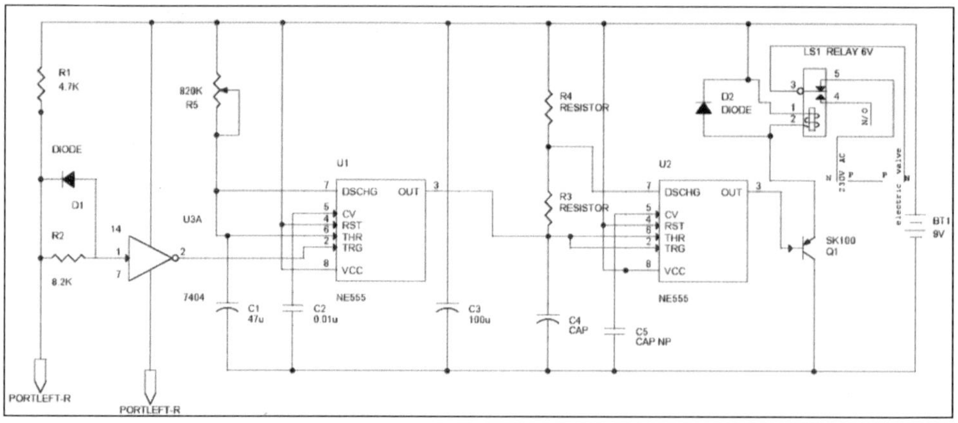

Plant Irrigation System Circuit Diagram

To conduct the probes, the voltage supply is provided from a battery, which is connected to the circuit. When the soil becomes dry, it produces large voltage drop due to high resistance, and this is sensed by the hex inverter and makes the first NE555 timer. This timer is arranged as a monostable multivibrator with the help of an electrical signal.

When the first 555 timer is activated at pin2, it generates the output at pin3; and, this output is given to the input of the second timer. This second NE555 timer is configured with astable multivibrator and generates the output to make the relay which is connected to the electrically operated value through the SK100 transistor. The output of the second timer switches on the transistor that drives the relay. This relay is connected to the input of an electrical value and the output of the electrical value is given to the plants through the pipe.

When the relay is turned on, the valve opens and water through the pipes rushes to the crops. When the water content in the soil increases, the soil resistance gets decreases and the transmission of the probes gets starts to make the inverter stop the triggering of the first timer. Finally the valve which is connected to the relay is stopped.

The advantage of using an automatic plant irrigator is that it is a simple system capable of conserving water, improving growth, discouraging weeds, saving time, and controlling fungal diseases and adaptable to the conditions.

Micro Irrigation System

Micro irrigation is nothing but a slow and regular application of water and nutrients moving down drop-by-drop directly to the root zone of the plants through low-discharge emitters and plastic pipes. This irrigation system is today's need of the hour as the natural water resources which are gift to the mankind have become scarce, and that are now not unlimited and free forever. But, the world's water resources are now fast moving back on track. After one completes the study of inter relationship between crops, soil, water and climatic conditions, one will find this micro irrigation system as a suitable system capable of delivering exact quantity of water at the root zone of the plants.

Micro Irrigation System

This system ensures that the plants do not endure from the strain or stress of less and over watering. The advantages of using this micro irrigation system are that for every drop of water used, we get more crop, better quality, early maturity, higher yield. Moreover, this system saves labor cost and water up to 70%. The working of this irrigation system covers over 40 crops spanning across 500 acres.

Automatic Plan Irrigation System using Microcontroller

Every irrigation system such as drip, sprinkler and surface gets automated with the help of electronic appliances and detectors such as computer, timers, sensors and other mechanical devices.

Automatic irrigation system

An automatic irrigation system does the work quite efficiently and with a positive impact on the place where it is installed. Once it is installed in the agricultural field, the water distribution to crops and nurseries becomes easy and doesn't require any human support to perform the operations permanently. Sometimes automatic irrigation can also be performed by using mechanical appliances such as clay pots or bottle irrigation system. It's very hard to implement irrigation systems because they are very expensive and complex in their design. By taking some basic points into considerations from experts' support, we have implemented some projects on automatic irrigation system by using different technologies.

Automatic Irrigation System on Sensing Soil Moisture Content

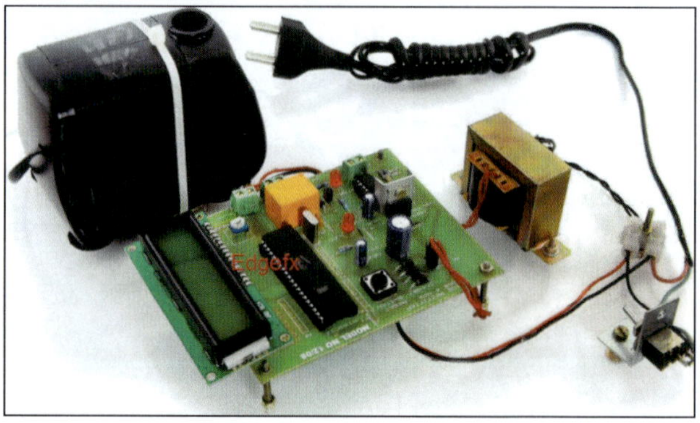

Automatic Irrigation System Circuit

The automatic irrigation system on sensing soil moisture project is intended for the development of an irrigation system that switches submersible pumps on or off by using relays to perform this action on sensing the moisture content of the soil. The main advantage of using this irrigation system is to reduce human interference and ensure proper irrigation.

The Microcontroller acts as a major block of the entire project, and a power supply block is used for supplying power of 5V to the whole circuit with the help of a transformer, a bridge rectifier circuit and a voltage regulator. The 8051 microcontroller is programmed in such a way that it receives the input signal from the sensing material which consists of a comparator to know the varying conditions of the moisture in the soil. The OP-AMP which is used as comparator acts as an interface between the sensing material and the microcontroller for transferring the moisture conditions of the soil, viz.wetness, dryness, etc.

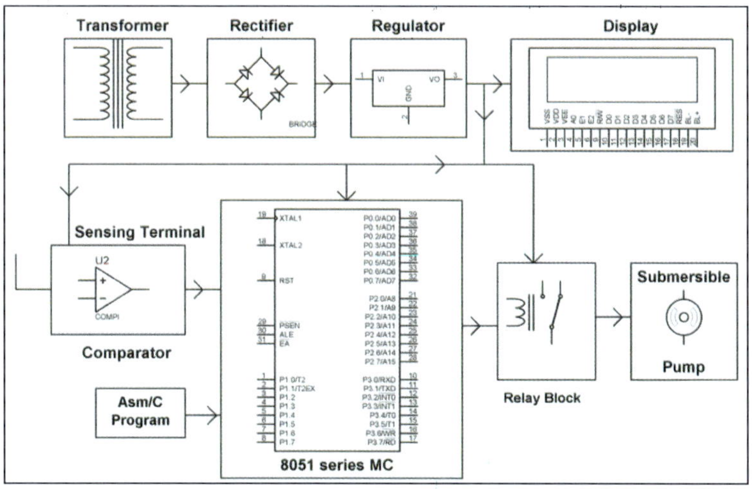

Block Diagram of Soil Moisture Content Based Irrigation

Once the microcontroller gets the data from the sensing material – it compares the data as programmed in a way, which generates output signals and activates the relays for operating the submersible pump. The sensing arrangement is done with the help of two stiff metallic rods that are inserted into the agricultural field at some distance. The required connections from these metallic rods are interfaced to the control unit for controlling the operations of the pump according to the soil moisture content.

This automatic irrigation system can be further enhanced by using advanced technology that consumes solar energy from solar panels.

Solar Powered Auto Irrigation System

Solar Powered Auto Irrigation System Circuit

In the above figure, the power from utilities is required to operate the system. As an extension to the above discussed system, this system uses solar panels to power the circuit. In agricultural field, the proper usage of automatic irrigation method is very vital due to some shortcomings of the real world like scarcity of land reservoir water and scarcity of rainfall. The water level (the ground water table) is getting reduced due to continuous extraction of water from the ground and thus gradually resulting in water scarcity in the agricultural zones slowly turning them into barren lands.

In the above irrigation system, solar energy generated from the solar panels is used for operating the irrigation pump. The circuit comprises moisture sensors built by using OP-AMP IC. The OP-AMP is used as comparators. Two stiff copper wires are inserted into the soil to know whether soil is wet or dry. A charge controller circuit is used to charge the photovoltaic cells for supplying the solar energy to the whole circuit.

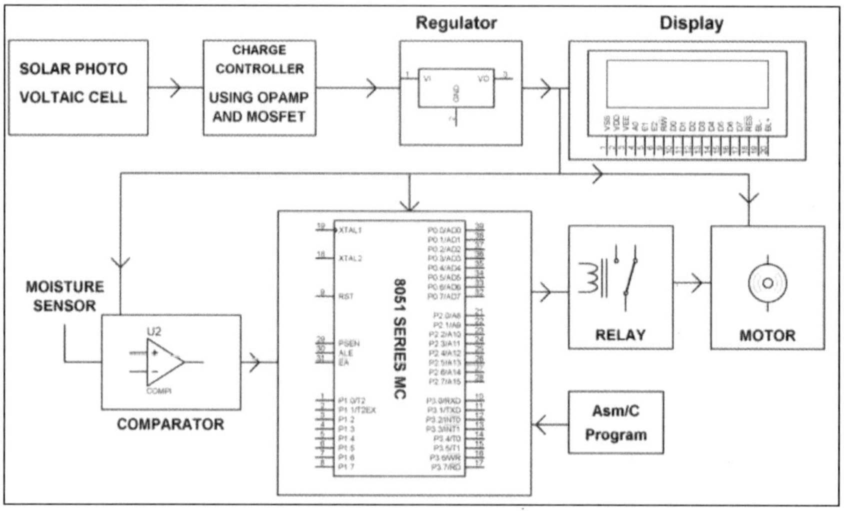

Solar Powered Auto Irrigation System Blockdiagram

A moisture sensor is used for sensing the soil condition – to know whether the soil is wet or dry, and the input signals are then sent to the 8051 microcontroller, which controls the whole circuit. The microcontroller is programmed by using KEIL software. Whenever the soil condition is 'dry', the microcontroller sends commands to the relay driver and the motor gets switched on and supplies water to the field. And, if the soil gets wet, the motor gets switched off.

The signals that are sent from the sensors to the microcontroller through the output of the comparator operate under the control of a software program which is stored in the ROM of the microcontroller. The LCD displays the condition of the pump (on or off) interfaced to the microcontroller.

This automatic irrigation system can be further enhanced by using GSM technology to gain control over the switching operation of the motor.

GSM based Automatic Irrigation System

Nowadays farmers are struggling hard in the agricultural fields round the clock. They do their field work in the morning section and irrigate their land during night time with intermittent intervals. The task of irrigating fields is becoming quite difficult for the farmers due to lack of regularity in their work and negligence on their part because sometimes they switch on the motor and then forget to switch off, which may lead to wastage of water. Similarly, they even forget to switch on the irrigation system, which again leads to damage to the crops. To overcome this problem, we have implemented a new technique by using GSM technology, which is explained below.

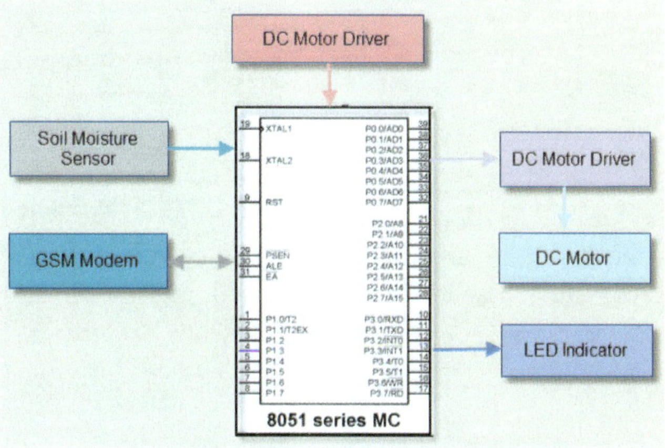

GSM Based Automatic Irrigation System

The GSM Based automatic irrigation system is a project in which we get update status of the operation carried out in the agricultural fields via SMS with the help of a GSM modem. We can also add other systems such as LCD displays, web cam and other smart controlled devices. In this project, we are using LEDs for indication purpose.

In this project, we are using soil moisture sensor which is used to sense the moisture level in the to know whether it is dry or wet. The moisture sensor is interfaced with the microcontroller. The input data signals from the moisture sensor are sent to the microcontroller and based on that it activates the DC Motor and switches the motor on with the help of a motor driver. After the soil gets wet, the Motor gets switched off automatically. The status of the agricultural fields can be known from the indication of the Light Emitting Diode (LED) or through the message sent to the GSM modem placed at the field. Simultaneously it is possible to send messages through a mobile to kit through the GSM modem. Thus, the irrigation motor can be controlled by using a mobile and a GSM modem.

References

- How-exactly-does-drip-irrigation-work: interestingengineering.com, Retrieved 30 March, 2019
- Irrigation-efficiency-types: aboutcivil.org, Retrieved 2 July, 2019
- Introduction-to-energy-efficient-irrigation: farm-energy.extension.org, Retrieved 1 April, 2019
- Ways-to-improve-irrigation-efficiency: agweb.com, Retrieved 4 June, 2019
- Soil-water-sensors-for-agriculture-theory-and-issues: conservationwebinars.net, Retrieved 5 April, 2019
- Spring-irrigation, agricultural-engineering: agritech.tnau.ac.in, Retrieved 3 February, 2019
- Irrigation-flow-meter: abestmeter.com, Retrieved 26 August, 2019
- Variable-rate-spraying: sprayers101.com, Retrieved 13 March, 2019
- Automatic-irrigation, irrigation, Soil-and-water, farm-management, agriculture: agriculture.vic.gov.au, Retrieved 1 January, 2019
- Automatic-plant-irrigation-system-circuit-and-its-working: edgefx.in, Retrieved 19 July, 2019
- Microcontroller-based-automatic-irrigation-system: elprocus.com, Retrieved 25 May, 2019

5

Harvesting and Postharvest Technology

The process of gathering ripe crops from the fields is known as harvesting. The stage which comes after harvesting is known as Postharvest. It includes various processes such as cooling, cleaning, sorting and packing. There are different types of harvesting machines such as potato harvesters and forage harvesters. The topics elaborated in this chapter will help in gaining a better perspective about the various technologies involved in harvesting and postharvesting.

Potato Harvester

At the time of harvesting, the harvesting process takes too much time hence potato harvesters are machines designed for potato harvesting Potato harvesters are machines designed for harvesting. This machine can finish mining, soil and potato separation in one time and can harvest the potato with the potato stems. It is a multifunctional machine. This is very efficient and also reduces damages of potatoes during harvesting.

Design

Components

Such development had been introduced to overcome the problems noticed under the harvesting operation using the ordinary digger, the digger unsuitable for harvesting root crops successfully, high percentage of losses as well as damage are resulted during the harvesting operation.

Frame

The frame is made of square pipe of 1.5" of mild steel. The frame takes a rectangular shape with dimension of 650 × 550 mm, include elements to fix (a spindle transports the vibrating movement to a cam at the end of it and vibrating system). The digger frame is carried by two tire wheels of 180 mm, diameter and 45 mm thickness

Separating Unit

The separating unit is consists of a blade with 545 × 200 × 6mm, which has 16 rods, 30 mm the distance between rods, this frame is connected to vibrating blade by nuts and bolts, also at the end of that frame longitudinal frame 500 × 450 × 10 mm is provided.

Transmission System

The transmission system consists of a spindle transmit rotating motion from to a pulley to the cam. The cam converts rotary motion into reciprocating motion tolinkage shaft. This shaft is connected through the longitudinal frame to digger blade.

Eccentric (Arm)

The main function of the arm is convert the rotating motion of the cam to reciprocating motion. Cam and follower are linked together by means of 2 bolts of M22.

Shaft

The shaft is made up of MS material having diameter of 30 mm. The diameter of the transmission shaft was calculated according to design of shaft for transmit the power considering the bending moment, axial load, and the torque acting on shaft.

Wheels

Wheels are used to support and carry the load of whole body of project. The wheels are used of diameter 200 mm for reduce height of project from ground level & project can work properly for underground root crops.

Working Principle

Potato harvester

This is simplest mechanism can used in project. The basic concept is that power transmission from one shaft to another shaft using two sprocket and chain but due to more noise and more cost and maintenance we can-not use this mechanism in this project.

The flat belt transmit very flow amount of power due to less friction of grip. It is require more space which increase the cost of project it give low velocity ratio while power transmit mission also the end of flat belt are joined so drive is not smooth.

Power Take Off (PTO)

Power Take Off shaft

PTO is any of several methods for taking power from power source such as running engine and transmitting it to an application such as an attached implement or separate machines.

Most commonly it is system comparising a splined output shaft on tractor, designed so that PTO shaft a kind of drive shaft can easily connected and disconnected, and a corresponding input shaft on application end. The PTO allows implement to draw energy from engine. The PTO and its associated shafts are couple by universal joints. The original type calls for operation at 540 RPM. A shaft that rotate at 540 rpm has 6 splines on it, having nominal diameter 35 mm. Agriculture PTOs are standardize in dimensions and speed.

Intermediate Shaft

Universal joint more commonly known as U joints, allow positive transmission of rotating power at much larger angle than is permissible with a flexible coupling.

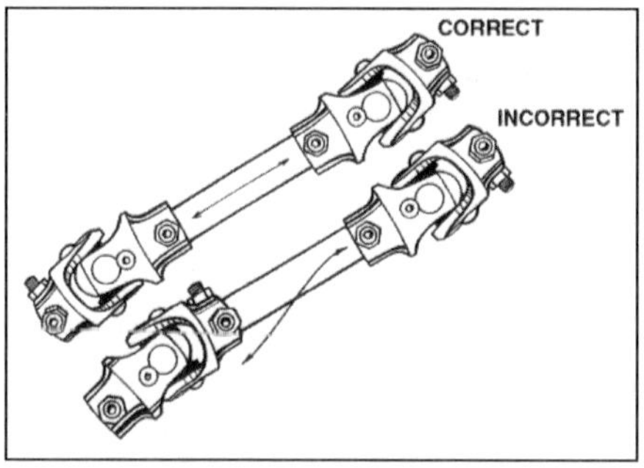

3 Intermediate Shaft

Millions of U joint are installed each year in all types of power transmission system thousands more are used to connect PTO drive shaft in off – highway tractors that operate drawn machinery such as rotary grass mowers, grain wagon with unloaders, feed grinder etc. Likewise, U joint widely used in industrial application U – joint used to join the drive shaft (PTO) and diffrential (machines).

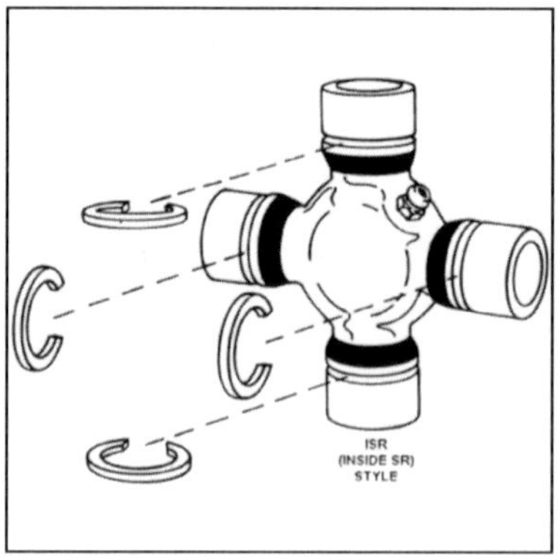

Universal joint

Advantages

1. This potato harvester machine can use for harvesting varieties of underground plants.
2. The machine is suitable for all kinds of soil, such as sandy soil, clay soil and roam.

3. The machine is with simple and compact structure and can connect with the tractors easily.
4. This potato harvesting equipment can harvest the potatoes with the potato vine.
5. Less man power require for operation.
6. It can use for multipurpose.
7. It saves the maximum time of operation than the manual.
8. It completes the operation without damaging the potatoes.

Applications

In majority of potato growing area digging is done with hand tools like khurpa, spade, plow and animal drawn plows. In recent years improvements have been made and multipurpose digger, digger shaker and digger windrower have been developed. So, for save the man power and time required for harvesting we design a suitable machine for easy to harvest root crops.

1. Potato Harvesting
2. Peanuts Harvesting
3. Sugar Beet Harvesting
4. Sweet Potato Harvesting.

Forage Harvesters

The flail type forage harvesters use free-swinging chains, hammers or knives to sever the fodder plants by beating or cutting action. At the time of plants being severed, the flails or knives travel in the same direction the machine is moving. The flail choppers do not have chopping knives to chop the material into acceptable lengths for silage. The flails are just used for severing the plants and harvested material can be blown into windrows for curing. The beating by the flails more or less conditions the hay.

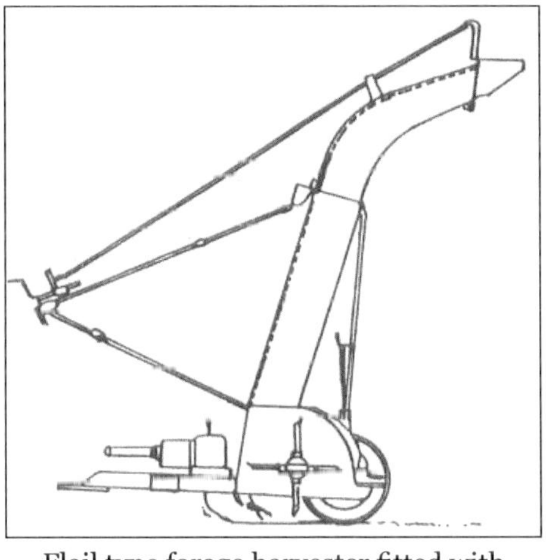

Flail type forage harvester fitted with adjustable twin-chop shear-plate

A simple flail type forage harvester can be mounted behind the tractor. Offset types of flail type forage harvester are preferred as they avoid problems associated with one set of tractor wheels running through the crop before it is cut. The degree of chopping and laceration is governed mainly by the rotor speed, partly by the relationship between rotor speed and forward speed and partly by the clearance between the flail tips and an adjustable shear bar. Rotor speed has much more effect than forward speed. Fitting two shear bars can reduce the length of chop. The machine can pick up crops fairly cleanly from a windrow.

A number of forage harvesters have large diameter cylinder with multiple knives, designed to avoid serious damage of cylinder by foreign materials, by allowing individual knife to either bend or forced inward to create wide clearance between knife and shear-plate. The damaged knife can be easily replaced. The machine has simplified feed and delivery systems. One such design uses a very wide contra-rotating cylinder in conjunction with a pair of feed rollers. The machine ejects the foreign matters such as small rocks and uses little energy in propulsion of the crop. Some new designs have achieved high throughput in relation to power input, partly as a result of less chopping. Electro-hydraulic systems are used on high-output forage harvesters for effective and rapid adjustment of discharge chute angle and the flaps.

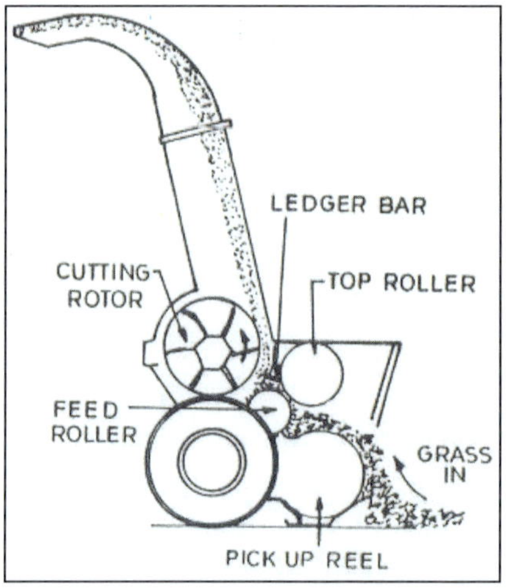

Forage harvester with pick-up, pair of feed roller and wide contra-rotating cylinder

Tractor Operated Flail Type Forage Harvester Cum Chopper

a) A view of forage harvester in field

This machine in a single operation can harvest chop and load the chopped fodder in the tractor-trailer attached to the machine. It is operated by a 26.1 kW tractor .This machine in a single operation can harvest, chop and load the chopped fodder like maize, bajra, and oats in the trailer attached to the machine. It consisted of a rotary shaft on which flails are mounted to harvest the crop, auger for conveying the cut crop, cutters for chopping & conveying chopped fodder through outlet into the trailer. After the blades cut the crop, it comes in auger, which conveys it to the chopping mechanism. The chopping mechanism cuts the crop into pieces and chopped material is thrown out with a high speed and is filled into the trailer hitched to the machine. Working capacity of forage harvester is 0.2 ha/h. Weight of the machine is about 670.0 kg.

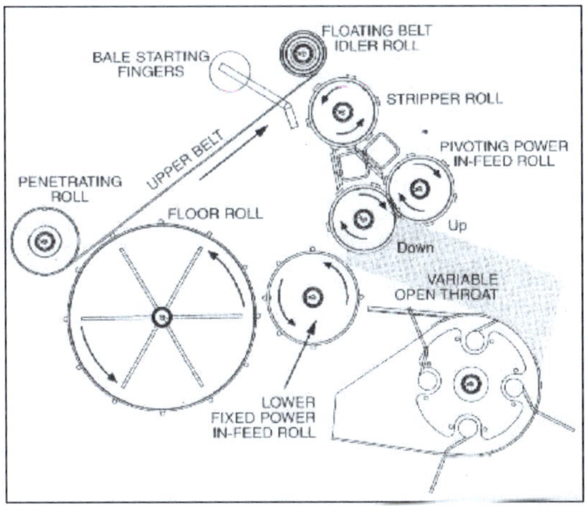

b) Details of forage harvester in field
Tractor operated Flail type forage harvester cum chopper

Tractor Operated Cutter Bar Type Forage Harvester and Loader

The system consisted of a two separate units. First one consisted of a mower and the second one was a chaff cutter cum loader. Both the machines are tractor operated. The mower has a working width of 196 cm and so designed that it can be folded easily during transportation. The machine cuts the fodder crops and lays in the field which is manually collected and then fed into the loader where it is chopped in to small pieces and directly loaded into the trailer. It is operated by a 26.1 kW tractor.

Tractor operated cutter bar type forage harvester and loader

The system consisted of a two separate units. First one consisted of a mower and the second one was a chaff cutter cum loader. Both the machines are tractor operated. The mower has a working width of 196 cm and so designed that it can be folded easily during transportation. The chaffer cum loader has a cross section area 28 x 13.5 cm^2. It has two powered feed roller and one compressing roller. It had two chaffing blades for chaffing of fodder and six thrower attachments are being mounted on the periphery of the cutter. A reversing mechanism has also provided for safety. The fodder cut with mover and is collected in the field manually. The chaffer-cum-loader chaps the fodder at site and the throwers mounted on the periphery of the cutter throws the fodder directly into the trailer. Working capacity 0.2 ha/h for cutter and 12-18 t/h for Chaffer-cum-loader. Weight of the machine (cutter and chaffer-cum-loader) is 150 + 690 kg. Overall dimensions are:

Self-propelled Fodder Harvester (Cutter Bar Type)

The self-propelled fodder harvester consists of 1450 mm cutter bar. It is mounted in the centre of the power transmission with the help of two side linkages made of high-pressure circular pipes to dampen the vibration of cutter bar. The unit is also provided with crop gathering drum-having flappers rotating inwards near the cutter bar. The vertical side covers around the transmission system are provided to avoid blockage of the fodder and extended vertical guides are provided for proper windrowing. The ground clearance of the machine has been increased to 480 mm to facilitate easy passage of fodder. Power to wheel is provided through extension with the help of chains and sprockets with the chassis suitably strengthened. The track width is also widened to 1040 mm and a caster wheel is provided at the rear to improve stability and maneuverability. The effective field capacity of machine is 0.1 ha/h at a forward speed of 1.5-2.0 km/h.

Self-employed fodder harvester (cutter bar type) in operation

Self-propelled Lucerne Harvester

Lucerne is perennial leguminous plant growing to a height of 60 to 90 cm. It is a forage crop generally grown in basins of size 1.20 x 6 m having a very high yielding potential under appropriate fertilizer and irrigation management. Harvesting of crop is one of the important agricultural operations, which demands considerable amount of labour because it is done manually by sickles. The cutting and laying in the windrows consume 65-75% of labour and gathering, bundle making and transport in the field involve the rest of the labour requirement. The scarcity and high cost of labour and drudgery during harvesting are the serious problem faced by the farmers. It is, therefore, essential to adopt the mechanical methods so that the timeliness in harvesting operation could be ensured and field losses are minimized to increase the productivity and production on the farm. The self-propelled Lucerne harvester consists of a gearbox and cutter bar. Type of cutter bar is bi-directional reciprocating type made from high carbon steel. Length of stroke for cutter bar is 25 mm and effective width of cutter bar is 860 mm. A man can walk behind the machine with an average speed of 2 km/h. The recommended speed ratio of the average cutter bar speed to the forward speed of machine is 1.3: 1.4. Two wheels are used for transportation purpose. The ground wheels drive the reel of the harvester. Ground drive provides the desirable feature of maintaining a constant speed ratio between peripheral speed and forward speed. The reel is made up from Ø 6 mm MS bar of diameter 70 cm and length 74 cm. The speed ratio of the ground wheel to reel is 1:1. The preliminary field trials were conducted at different locations. The effective field capacity was found to be 0.113 ha/h and field efficiency was 70-75%. Thus, there is net saving of 52% in cost of cultivation and net time saving of 90%.

A self–propelled lucerne harvester in operation

Self-propelled Cutter Bar Type Harvester for Fodder Crops

The machine consisted of a 10.2 hp, 4 stroke, single cylinder, air-cooled diesel engine. The machine has cutter bar width of 130 cm. The power from engine has been provided to the traction wheels with the help of gear trains. The machine has four forward speeds and one reverse. The track width of machine is 1.23 m and the ground clearance of machine 41 cm. The steering of machine is done with the help of foot-operated pedal provided below the operator seat, which guides the wheel. The crop after harvesting falls behind the cutter bar and was guided with wooden sticks provided behind the cutter bar for small height/hard stem crops. But in case of tall and weak stem crops, binding attachment without twine is required to avoid wrapping of crop to cutter bar. The average field capacity of the machine for harvesting barseem and oat crop was 0.42 and 0.28 ha/h respectively, whereas the corresponding throughput of material was 6.0 t/h and 11.0 t/h. The field efficiency of machine was 66.57 % and 40.55 % for barseem and oat crop. The height of cut from the ground for barseem and oat crop was 7.0 cm and 12.5 cm but it was more near the bunds. The saving in cost of operation and labour for harvesting barseem and oat crop with machine as compared to manual harvesting was 42-44 and 55% respectively.

Self-propelled cutter bar type harvester in operation for harvesting barseems and bajra folder

Hay Conditioners

The conditioning of hay by means of crushing, crimping or flailing is becoming increasingly popular with many hay-makers. It has many advantages viz. speeds up field curing and reduces drying time by 30%, reduces weather damage and field losses and conserves colour and feed value through shorter exposure and less shattering. There are three types of hay conditioners: smooth roll, corrugated roll or crisper and flail type forage harvester. The smooth rolls give continuous crushing action to hay, leaving no part uncrushed. It has two rolls of a combination of one rubber roll and another steel roll. Most rubber rolls have spiral grooves to aid in picking up the hay and feeding between the rolls. The corrugated-roll type hay conditioner is also equipped with two malleable iron rolls with tapered flutes that mesh together. Some conditioners use a fluted bar roll that presses against a smooth rubber roll. As the hay passes between the rolls, it is bent, crimped, cracked at intervals and in some cases crushed. The flail-type conditioner is a hay harvester, but it is used as a hay conditioner also. In this the shear bar is removed to reduce the cutting action. The hay is partially chopped by the swinging hammers or knives.

1. Separate conditioners: Separate conditioners usually pickup mowed hay from the swath, but windrowers can be put through them. The bottom roll acts as the pickup. Mower –conditioners with not over 2.75 m cuts have conditioner rolls with lengths approximately equal to the width of cut, and the hay is fed directly to the rolls from the cutter bar. Originally, conditioning ease often a separate operation, performed with a second tractor following the mower. Subsequently, mowers were made available with hitches and PTO extensions to permit pulling a conditioner behind the mower. The conditioner than picks up the swath adjacent to the one being cut.
2. Mower conditioners: Pull-type mower conditioner consists of a cutter bar, a reel, a pair of full-width conditioning rolls, and a deflector. The most common cutting width is 2.75 m, but widths range from 2.13 to 3.65 m. The 3.6 model have rolls 2.13 to 2.75 m long with short, auger-type cross conveyors at the outer ends of the header. The conditioned hay can be left evenly spread in a swath a little narrower that the cut or side deflectors behind the rolls can be adjusted to obtain various windrows widths and positions. Windrows cure more slowly that hay in the swath but can be straddled with the tractor and sometimes eliminate the raking operation. The cutter bar is similar to that of a conventional mower except that the guards are more pointed and slender to improve performance in tangled or lodged crops and are made in pairs to provide greater support in mounting.

Fruit Picking Robots

The robotics designers offer to the farmers the opportunity to significantly reduce the costs of manual labor for harvesting. The robots can replace the seasonal manual work or even permanent employees on farms.

Apple-picking Robot with Vacuum

Apple orchard farmers will be able to use robots instead of seasonal pickers. The AR startup uses the vacuum to pick apples from trees.

The robot uses computer vision algorithms to identify and locate apples in the tree. The technology used is not specifically designed for agriculture. The same technology can be applied in a wide range of industries, but for now they are using it into the agriculture.

Apples require attention at harvesting. The robot is designed to work with precision in harvesting and to store the apples. The collection is made through a flexible hose and the storage is made in the same big boxes as used by the human workers.

Autonomous Robot to Harvests Soft Fruits

Autonomous robot to harvests soft fruits

Strawberries are delicate fruits and require careful picking. The robot uses machine vision and motion planning algorithms to recognize and locate the ripe fruit to be picked.

The next step in robot development is to improve the learning algorithms. In order to make better fruit harvesting decisions, the robot needs algorithms capable of learning so that harvesting to be done with fewer errors. The fruit images are captured by several cameras that move up and down for a detailed view of the crops.

The GPS system mounted on the robot platform helps to precisely establish plants and fruit production. Thus, the farmer can identify the most productive area as well as the area with a low yield.

Robotic Fruit Harvester

Robotic Fruit Harvester

The robot is called FFRobot. Like many other robots, FFrobot uses advanced image processing algorithms and algorithms for picking the fruits from the tree. The image processing algorithms can detect damaged, diseased or unripe fruits.

The grasping hand can be easily modified to pick different types of fruits. This makes the robot available for different harvesting seasons.

SW 6010 to Harvest Strawberries

SW 6010 is the first autonomous robot available on the market able to detect and collect strawberries. The robot has attached a number of mobile arms that can identify and pick up the ripe strawberries. The robot analyzes the strawberries one by one and gently harvests them to prevent damage to the fruit.

SW 6010 to harvest strawberries

AGvision is a system developed by the company and uses artificial intelligence to identify the fruits and their quality. It can autonomously navigate between the rows of strawberries, so the operator can handle strawberries into boxes.

Combine Harvester

The combine harvester, or simply combine, is a machine that combines the tasks of harvesting, threshing, and cleaning grain crops.

A Lely open-cab combines

The objective is the harvest of the crop; corn (maize), soybeans, flax (linseed), oats, wheat, or rye, among others. The waste straw left behind on the field is the remaining dried stems and leaves of the crop with limited nutrients which is either chopped and spread on the field or baled for feed and bedding for livestock.

Combine Heads

Combines are equipped with removable heads (called headers) that are designed for particular crops. The standard header, sometimes called a grain platform (or platform header), is equipped with a reciprocating knife cutter bar, and features a revolving reel with metal or plastic teeth to cause the cut crop to fall into the head. A cross auger then pulls the crop into the throat. The grain header is used for many crops, including grain, legumes, and many other crops.

Wheat headers are similar except that the reel is not equipped with teeth. Some wheat headers, called "draper" headers, use a fabric or rubber apron instead of a cross auger. Draper headers allow faster feeding than cross augers, leading to higher throughputs. In high yielding European crops, such headers have no advantage, as the limiting factor becomes grain separation. On many farms, platform headers are used to cut wheat, instead of separate wheat headers, so as to reduce overall costs.

Dummy heads, or pick-up headers, feature spring-tined pickups, usually attached to a heavy rubber belt. They are used for crops that have already been cut and placed in windrows or swaths. This is particularly useful in northern climates, such as western Canada, where swathing kills weeds, resulting in a faster dry down.

A John Deere 9410 Combine set to harvest Oats

While a grain platform can be used for corn, a specialized corn head is ordinarily used instead. The corn head is equipped with snap rolls that strip the stalk and leaf away from the ear, so that only the ear (and husk) enters the throat. This improves efficiency dramatically since so much less material must go through the cylinder. The corn head can be recognized by the presence of points between each row.

Occasionally, row-crop heads are seen that function like a grain platform, but have points between rows like a corn head. These are used to reduce the amount of weed seed picked up when harvesting small grains.

A combine harvesting corn

Self-propelled Gleaner combines could be fitted with special tracks instead of tires to assist in harvesting rice. Some combines, particularly pull type, have tires with a diamond tread, which prevents sinking in mud.

Conventional Combine

The cut crop is carried up the feeder throat by a "chain and flight elevator," then fed into the threshing mechanism of the combine, consisting of a rotating threshing drum, to which grooved steel bars are bolted. These bars thresh or separate the grains and chaff from the straw through the action of the drum against the concave, a shaped "half drum," also fitted with steel bars and a meshed grill, through which grain, chaff and smaller debris may fall, whereas the straw, being too long, is carried through onto the straw walkers. The drum speed is variably adjustable, whilst the distance between the drum and concave is finely adjustable fore, aft and together, to achieve optimum separation and output. Manually engaged disawning plates are usually fitted to the concave. These provide extra friction to remove the awns from barley crops.

Sidehill Leveling

An interesting technology is in use in the Palouse region of the Pacific Northwest of the United States, in which the combine is retrofitted with a hydraulic sidehill leveling system. This allows the combine to harvest the incredibly steep but fertile soil in the region. Hillsides can be as steep as a 50 percent slope. Gleaner, IH and Case IH, John Deere, and others all have made combines with this sidehill leveling system, and local machine shops have fabricated them as an after-market add-on.

The first leveling technology was developed by Holt Co., a California firm, in 1891. Modern leveling came into being with the invention and patent of a level sensitive mercury switch system invented by Raymond Hanson in 1946. Raymond's son, Raymond, Jr., produced leveling systems exclusively for John Deere combines until 1995, as R. A. Hanson Company, Inc. In 1995, his son, Richard, purchased the company from his father and renamed it RAHCO International, Inc. In April, 2007, the company was renamed The Factory Company International, Inc. Production continues to this day.

Sidehill leveling has several advantages- Primary among them is an increased threshing efficiency on sidehills. Without leveling, grain and chaff slide to one side of separator and come through the machine in a large ball rather than being separated, dumping large amounts of grain on the ground. By keeping the machinery level, the straw-walker is able to operate more efficiently, making for more efficient threshing. IH produced the 453 combine, which leveled both side-to-side and front-to-back, enabling efficient threshing whether on a sidehill or climbing a hill head on.

Secondarily, leveling changes a combine's center of gravity relative to the hill and allows the combine to harvest along the contour of a hill without tipping, a very real danger on the steeper slopes of the region; it is not uncommon for combines to roll on extremely steep hills.

Currently, sidehill leveling is on the decline with the advent of huge modern machines which are more stable due to their width. These modern combines use the rotary grain separator which makes leveling less critical. Most combines on the Palouse have dual drive wheels on each side to stabilize them.

Maintaining Threshing Speed

Another technology that is sometimes used on combines is a continuously variable transmission. This allows the ground speed of the machine to be varied while maintaining a constant engine and threshing speed. It is desirable to keep the threshing speed since the machine will typically have been adjusted to operate best at a certain speed.

Self-propelled combines started with standard manual transmissions that provided one speed based on input rpm. Deficiencies were noted, and in the early 1950s, combines were equipped with what John Deere called the "Variable Speed Drive." This was simply a variable width shive controlled by spring and hydraulic pressures. This shive was attached to the input shaft of the transmission. A standard 4 speed manual transmissions was still used in this drive system. The operator would select a gear, typically third. An extra control was provided to the operator to allow him to speed up and slow down the machine within the limits provided by the variable speed drive system. By decreasing the width of the shive on the input shaft of the transmission, the belt would ride higher in the groove. This slowed the rotating speed on the input shaft of the transmission, thus slowing the ground speed for that gear. A clutch was still provided to allow the operator to stop the machine and change transmission gears.

Allis-Chalmers GLEANER L2

Later, as hydraulic technology improved, hydrostatic transmissions were introduced by Versatile Mfg for use on swathers but later this technology was applied to combines as well. This drive retained the 4 speed manual transmission as before, but this time used a system of hydraulic pumps and motors to drive the input shaft of the transmission. This system is called a Hydrostatic drive system. The engine turns the hydraulic pump capable of high flow rates at up to 4000 psi. This pressure is then directed to the hydraulic motor that is connected to the input shaft of the transmission. The operator is provided with a lever in the cab that allows for the control of the hydraulic motors ability to use the energy provided by the pump. By adjusting the swash plate in the motor, the strokes of its pistons are changed. If the swash plate is set to neutral, the pistons do not move in their bores and no rotation is allowed, thus the machine does not move. By moving the lever, the swash plate moves its attached pistons forward, thus allowing them to move within the bore and causing the motor to turn. This provides an infinitely variable speed control from 0 ground speed to whatever the maximum speed is allowed by the gear selection of the transmission. The standard clutch was removed from this drive system, as it was no longer needed.

Most, if not all, modern combines are equipped with hydrostatic drives. These are larger versions of the same system used in consumer and commercial lawn mowers that most are familiar with today. In fact, it was the downsizing of the combine drive system that placed these drives systems into mowers and other machines.

Claas Lexion 570

The Threshing Process

Despite great advances mechanically and in computer control, the basic operation of the combine harvester has remained unchanged almost since it was invented.

First of all the header, cuts the crop and feeds it into the threshing cylinder. This consists of a series of horizontal rasp bars fixed across the path of the crop and in the shape of a quarter cylinders, guiding the crop upwards through a 90 degree turn. Moving rasp bars or rub bars pull the crop through concaved grates that separate the grain and chaff from the straw. The grain heads fall through the fixed concaves onto the sieves. The straw exits the top of the concave onto the straw walkers.

Since the IH 1440 and 1460 Axial-Flow Combines came out in 1977, combines have rotors in place of conventional cylinders. A rotor is a long, longitudinally mounted rotating cylinder with plates similar to rub bars.

There are usually two sieves, one above the other. Each is a flat metal plate with holes set according to the size of the grain mounted at an angle which shakes. The holes in the top sieve are set larger than the holes in the bottom sieve. While straw is carried to the rear, crop and weed seeds, as well as chaff, fall onto the second sieves, where chaff and crop fall though and are blown out by a fan. The crop is carried to the elevator which carries it into the hopper. Setting the concave clearance, fan speed, and sieve size is critical to ensure that the crop is threshed properly, the grain is clean of debris, and that all of the grain entering the machine reaches the grain tank. (For example, when traveling uphill, the fan speed must be reduced to account for the shallower gradient of the sieves).

Heavy material, such as unthreshed heads, fall off the front of the sieves and are returned to the concave for re-threshing.

The straw walkers are located above the sieves, and also have holes in them. Any grain remaining attached to the straw is shaken off and falls onto the top sieve.

When the straw reaches the end of the walkers it falls out the rear of the combine. It can then be baled for cattle bedding or spread by two rotating straw spreaders with rubber arms. Most modern combines are equipped with a straw spreader.

Rotary vs. Conventional Design

For a considerable time, combine harvesters used the conventional design, which used a rotating cylinder at the front-end which knocked the seeds out of the heads, and then used the rest of the machine to separate the straw from the chaff, and the chaff from the grain.

Case IH Combine set to harvest soybeans

In the decades before the widespread adoption of the rotary combine in the late seventies, several inventors had pioneered designs which relied more on centrifugal force for grain separation and less on gravity alone. By the early eighties, most major manufacturers had settled on a "walkerless" design with much larger threshing cylinders to do most of the work. Advantages were faster grain harvesting and gentler treatment of fragile seeds, which were often cracked by the faster rotational speeds of conventional combine threshing cylinders.

The disadvantages of the rotary combine (which were increased power requirements and pulverization of the straw by-product) prompted a resurgence of conventional combines in the late nineties. Perhaps overlooked, but nonetheless true, when the large engines utilized to power the rotary machines were employed in conventional machines, the two types of machines delivered similar production capacities. Also, research was beginning to show that incorporating above-ground crop residue (straw) into the soil is less useful for rebuilding soil fertility than previously believed. This meant that working pulverized straw into the soil became more of a hindrance than a benefit. An increase in feedlot beef production also created a higher demand for straw as fodder. Conventional combines, which use straw walkers, preserve the quality of straw and allow it to be baled and removed from the field.

Threshing Equipment

Features

The thresher consists of rasp bar threshing cylinder, oscillating sieves, concave and winnowing and cleaning attachment. The rasp-bar cylinder is made of cast iron rings, sheet metal and toothed racks and is fitted on the main shaft running on two heavy pedal bearings. Various pulleys of different sizes are fitted on this shaft to transmit power to winnowing and fan attachment. Adjustments for cylinder and blower speeds and concave clearance are provided to make the machines suitable for threshing various crops. It has a safe feeding chute attached to it.

Specification

Diameter of beater shaft (mm)	42
Length of beater shaft (mm)	1400
Diameter of cylinder with rasp-bar (mm)	416
Length of the cylinder (mm)	560
Height of hopper from ground level (mm)	1500
Capacity (kg/h)	600-1000
Power requirement (hp/kW)	7.5/5.6

Uses: It is suitable for threshing paddy, ragi, jowar, maize, sunflower, wheat, mustard etc.

Axial-Flow Paddy Thresher

Features

It consists of a threshing cylinder, concave, cylinder casing, cleaning system and feeding chute. In axial flow concept, the crop is fed from one end, it moves axially and the straw is thrown out from the other end after complete threshing of the crop. During threshing, the crop rotates three and half times around the cylinder and all the grains get detached. The threshing cylinder is of peg type. The casing of the thresher has 7 louvers for moving.

Specifications

Length (mm)	2030
Width (mm)	3050
Height (mm)	1960

Feeding device: Length of feeding chute (mm) Width of feeding chute (mm)	Feeding chute, manually 905 110
Type of cylinder	Spike tooth
Cylinder size, tip diameter x length (mm)	770 × 1500
Length of concave (mm)	840
Width of concave (mm)	570
Concave clearance (mm)	18-21
Type, size and number of blowers	Aspirator type, two blowers of 400 mm diameter and 160 mm width
Weight (kg)	730
Power requirement(hp/Kw)	35/26.25, tractor PTO

Uses: It is used for threshing of paddy crop.

Pedal Operated Paddy Thresher

Features

It consists of wire-loop type threshing cylinder, power transmission system, mild steel sheet body and foot pedal. The threshing cylinder consists of wire-loops of 'U' shape embedded in wooden or metallic strips joined to two discs. A shaft carries the threshing cylinder and is connected to the transmission system. The transmission system consists of meshed gears or sprocket-chain mechanism. The larger gear or sprocket is connected to foot pedal/bar with links. The foot pedal/bar is always in raised position. On pressing the pedal the threshing cylinder starts rotating. For continuous rotation of the cylinder, the pedal is lowered and raised repeatedly. For operation, paddy bundle is held in hands and earhead portion of the crop is placed on the rotating cylinder. The wire-loops hit the earheads and grain get detached from the rest of the crop.

Specifications

Dimensions (LxWxH) (mm)	1250 x 650 x 630
Cylinder size (mm)	400 diameter x 600 length

Loop wire diameter (mm)	4
Number of wire loops	44
Number of wire loop strips	12
Transmission system	Pedal Crank mechanism with 1:4 gear ratio
Weight (kg)	36
Power source	One Person

Uses: It is used for threshing of paddy crop. It easy to operate and does not require much effort in its operation as it is operated by foot.

Maize De-husker/Sheller

Features

Maize de-husker cum sheller are of two types namely, spike tooth type and axial flow type. In the spike tooth type sheller, pegs are staggered at varying heights for better shelling efficiency. The spikes are placed in 6 rows with 6 spikes in each row. The sieves have 1.25 cm diameter opening to separate the shelled maize from husk. In axial flow type threshers, pegs are provided on the cylinder and louvers were provided on the upper periphery of the drum to convey the crop to the outlet.

Specification

	Spike tooth type	Axial type
Length (mm)	1765	3400
Width (mm)	1065	1140
Height (mm)	1570	2190
Dia. of threshing cylinder (mm)	480	495
Length of threshing cylinder (mm)	325	1460
Cylinder speed (m/s)	16.4	13.7
Type of threshing drum	Spike tooth	Peg type
Type of blower	Centrifugal	Centrifugal

No of sieves	2	3
Opening size of upper sieve (mm)	12	12.5
Opening size of middle sieve (mm)	-	7
Opening size of lower sieve (mm)	2	3
Concave clearance (mm)	50	50
Opening size of concave (mm)	5 x 5	5.1 x 5.1
Moisture content of cobs (%)	12-28	12-28
Threshing Capacity (kg/hr)	450-650	1200-2800
Power source (hp/kW)	35/26, Tractor	35/26, Tractor

Uses: It is used for threshing of maize cobs.

Power Thresher

It is a machine operated by a prime mover such as electric motor, engine, and tractor or power tiller used for threshing.

A power thresher performs several functions such as:

1. To feed the harvest to the threshing cylinder.
2. To thresh the grain out of the head.
3. To separate the grain from the straw.
4. To clean the grain.
5. To put the grain in a bag.
6. To make chaff, suitable for animal feeding.

Removal of seeds from the grain heads is done by rotating cylinders, whose threshing action depends primarily upon impact. When a slow moving material comes in contact with the high speed cylinder, the heads or pods are shattered and grains arc freed from straw. Further threshing is done when the material passes through the restricted clearance space between the thresher cylinder and the concave portion of the unit.

Types of Power Thresher

There are following types of thresher:

1. Hammer mill type.
2. Rasp bar cylinder type.
3. Spike-tooth cylinder type.
4. Syndicator type.
5. Drummy type thresher.

1. Hammer Mill Type: It is a thresher with threshing unit consisting of hammers or beaters with a closed cylinder casing and concave. It is equipped with a set of oscillating sieves and an aspiratory blower for separation and cleaning of grains.

2. Rasp-Bar Cylinder Type: In this thresher, the thresher unit consists of bars with serrations having an open concave.

3. Spike Tooth Cylinder Type: It is a thresher, the threshing unit of which consists of drum having rows of spikes with a closed cylinder casing and concave and equipped with a set of sieves and aspiratory blower.

4. Syndicator Type: It is a thresher, the threshing unit of which consists of a corrugated flywheel with serrated chopping knives and a closed cylinder casing and concave. This is also known as chaff-cutter type thresher.

5. Drummy Type Thresher: It is a hammer mill type thresher without separation and cleaning system. Usually a centrifugal blower is provided for partial separation and cleaning of grain.

On the basis of feeding system, the power thresher can be of four types:

1. Chute Feed Thresher: A thresher in which the feeding of the crop is done through a chute.
2. Conveyer Feed Thresher: A thresher in which the feeding of crop is done through a conveyer.
3. Feed Roller Feed Thresher: A thresher in which the feeding is done with the feed rollers equipped with chute or an endless conveyer.
4. Hopper Feed Thresher: A thresher in which feeding of the crop is done through the hopper. It is also known as bulk feed thresher.

On the basis of crop, thresher may be of following types:

1. Wheat Thresher: This equipment is used for threshing of wheat crop with or without chaff making provision.
2. Paddy Thresher: This equipment is used for threshing paddy crop.
3. Groundnut Thresher: This equipment is used for threshing of groundnut.
4. Millet Thresher: This equipment is used for threshing of millet crop.
5. Soya Bean Thresher: This equipment is used for threshing of soya bean crop.
6. Multi Crop Thresher: This equipment is used for more than one crop with or without minor adjustment.

This thresher has either spike tooth cylinder or rasp bar cylinder depending upon the manufacturer. It has cleaning and bagging attachments. This thresher can be used for crops like paddy, wheat, sorghum, soya bean, gram, millets etc. It can be operated by 5-20 hp power depending upon the models. Its capacity may be 300-2500 kg/hr.

Components of Power Thresher

The main components are:

1. Concave
2. Cylinder or drum
3. Cleaning unit.

Concave

It is a concave shaped metal grating, partly surrounding the cylinder against which the cylinder rubs the grain from the plant or ear heads and through which the grains fall on the sieve.

Cylinder or Drum

It is a balanced rotating assembly, comprising rasp, beater bar or spikes on its periphery and their support for threshing the crop.

There are five types of threshing cylinders commonly used in the country:

1. Peg Tooth Cylinder: The teeth on the concave and cylinder are so arranged that the cylinder teeth pass midway between the staggered teeth on the concave. The concave assembly is pivoted at the rear portion of the machine. The clearance space between the cylinder and the concave is adjusted according to the requirement. As the stalks pass through the clearance space, the grains get separated from the head due to impact action between the teeth.

2. Loop Type: The cylinder is studded with a number of wire loops throughout its outer periphery. This is mostly used on paddy threshers.

3. Angle Bar Cylinder: Cylinder is equipped with angle iron bars, helically fitted on the cylinders. The bars have rubber pads on their faces. The concave unit is fitted with a rubber faced shelling plate and steel jacketed rubber bars. The clearance between the cylinder and concave unit at the entrance is from 13 mm to 19 mm and reduces to about 6 to 9 mm only.

4. Hammer Mill Type: The beaters are in the shape of hammer mill. The beaters are attached with the beater arm at the tip. Beater arms are rigidly fixed to a hub which is mounted on main shaft.

5. Rasp Bar Cylinder: The cylinder has corrugated bars round it. Threshing is accomplished between corrugated cylinder bars and stationary bars of the concave portion. The rotating cylinder takes the grains out from the head as it is drawn over, the bars on the concave unit. Usually 6 to 8 bars are spirally fixed on the cylinder.

Cleaning Unit

The function of the cleaning unit is to separate and clean the threshed grain. The cleaning unit mainly consists of two or more oscillating sieves, a fan and an air sucking duct known as aspirator.

Usually two ducts are there, one primary duct and the other secondary duct. The function of the primary duct is to remove major portion of straw, dust and other foreign matter. The secondary duct is used for final cleaning of the grains.

Thresher provided with aspirator unit is usually called aspirator type of thresher. Those threshers which are not fitted with aspirator unit have got only one blower, which blows air in horizontal direction. This type of thresher is commonly called drummy thresher:

1. Aspirator – It is a component of the cleaning unit used for cleaning grain by drawing air through the grain mass.
2. Blower – It is a device to produce air blast.
3. Winnower – It is a machine with one or two sieves and fan using air. Drummy thresher stream across falling grain.
4. Winnowing Fan – It is a machine used for creating air blast mainly for the purpose of winnowing of grains.
5. Seed Damage – Seed damage may occur due to cylinder concave clearance being too small. In some cases the damage is due to the impact blow which is directly related to the cylinder peripheral speed. The speed damage may or may not be visible. The internal damage may be known only by germination test.
6. Cylinder Adjustment – Cylinder concave clearance may be adjusted by raising or lowering the cylinder and the concave unit. Clearance should be as great as can be used with satisfactory threshing. Cylinder speeds may be changed by changing sheaves and sprockets.

Installation of Power Thresher

1. The prime mover of thresher should be located such that blown straw does not fall on the thresher.
2. The prime mover should be kept at a safe distance from the feeding mechanism.
3. The prime mover should be suitably levelled.
4. Straw outlet should be kept in the direction of wind.
5. The power thresher should be firmly grounded for stability.
6. Proper alignment of thresher should be done.
7. Direction of rotation of belt should be properly maintained.
8. The tension of the driving belt should be maintained.
9. Guards and safety devices should be maintained. Proper feeding should be done for safety of operators.

Recommended Speed of Threshing Drum

Crop	Speed of Drum	
	rev/min	metre/second
Paddy	675-1000	16 to 25
Wheat	550-1100	20 to 30
Barley	740-1080	20 to 26
Gram	400-750	12 to 22
Jawar	400-675	12 to 20
Bajra	400-550	12 to 16
Peas	430-750	12 to 22

Diameter of thresher pulley (D_t) is given by:

$$D_t = D_p \times N_p / N_t$$

Where,

D_t = diameter of the thresher pulley (mm),

D_p = diameter of the prime mover pulley (mm),

N_p = speed of the prime mover pulley (rev/min),

N_t = speed of thresher pulley (rev/min).

Maintenance and Storage of Power Thresher

Preventive Maintenance of Power Thresher:

1. Belts to be checked.
2. Bolts and screws to be checked.
3. Foreign matter (if any) in the crop should be removed.
4. Damaged and bent parts to be checked and repaired.
5. Sieves to be checked and cleaned if openings are clogged.
6. Point to be cleaned and greased.

Storage of Thresher in Off-season

1. Run the thresher idle for some time after the work is over.
2. Disconnect the power source and remove all the grains and straw.
3. Remove all belts, clean and store them in a safe place.
4. Wash, clean and dry the thresher completely.
5. Repaint the thresher where necessary or apply a coat of grease or used engine oil.

6. Lubricating points such as grease cups and bearings should be cleaned with kerosene oil or diesel oil and should be lubricated with fresh oil or grease.
7. Entrance of virmins should be prohibited by closing the passage.
8. The thresher should be stored in dry shed.
9. Jack the thresher frame horizontally on wooden blocks or bricks in case of pneumatic tyres.

Trouble Shooting in Power Thresher

Some of the reasons and remedies are:

Trouble – 1- Threshing Drum is Blocked

1. Reason – Over feeding. Remedy – Avoid over feeding.
2. Reason – Belt loose. Remedy – Check all the belt.
3. Reason -Drum speed too low. Remedy – Increase the speed.
4. Reason – Moist crop. Remedy – Dry the crop.
5. Reason – Crop infested with weeds. Remedy- Remove the weeds.
6. Reason – Less concave clearance. Remedy – Adjust the proper clearance.

Trouble – 2 – Broken Grains Coming out of Thresher

1. Reason – High speed of drum. Remedy – Decrease the drum speed.
2. Reason – Less concave clearance. Remedy – Adjust the proper clearance.

Trouble – 3 – Grains Blows with Straw

1. Reason – High speed of fan. Remedy – Decrease the fan speed.
2. Reason – Sieve holes blocked. Remedy – Clean the sieve holes.

Trouble- 4 – Straw is coming with Grain

1. Reason – Low fan speed. Remedy – Adjust the proper fan speed.
2. Reason – Improper size of sieves. Remedy – Use proper size of sieves.
3. Reason – Too much feeding and uneven feeding. Remedy – Proper feeding should be done.
4. Reason – Blocked concave. Remedy – Clean the concave.
5. Reason – Low shaking speed. Remedy – Adjust the proper shaking speed.

Trouble – 5 – Grain is coming with Tailing

1. Reason – Upper sieve blocked. Remedy – Clean upper sieve.

2. Reason – Improper sieve slope and size. Remedy – Adjust proper sieve slope and use proper size of sieve.

3. Reason – High shaking speed. Remedy – Decrease shaking speed.

Trouble – 6 – Too Many Unthreshed Heads

1. Reason – Low cylinder speed. Remedy – Increase cylinder speed.

2. Reason – High concave clearance. Remedy – Decrease concave clearance.

Trouble – 7 – Vibration in Thresher

1. Reason – Improper installation. Remedy – Install the thresher properly.

2. Reason – Bearings are loose. Remedy – Tighten the bearing.

3. Reason – Worn-out components. Remedy – Repair or replace the components.

Drying Methods

The below figure shows the detailed classification of drying methods, which are of relevance to regions of world.

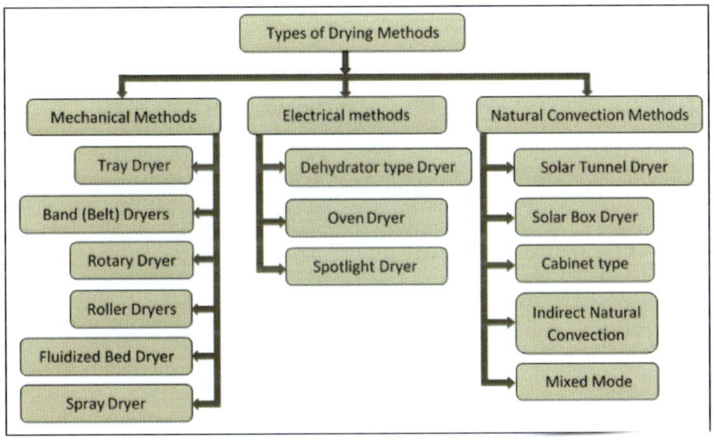

Drying methods

Mechanical Methods

Tray Dryer

Tray dryers are classified as batch type and band dryer and can dry almost everything. However, because of the labors required for loading and unloading, they are expensive to operate. They find most frequent application when valuable products like dyes and pharmaceuticals are involved.

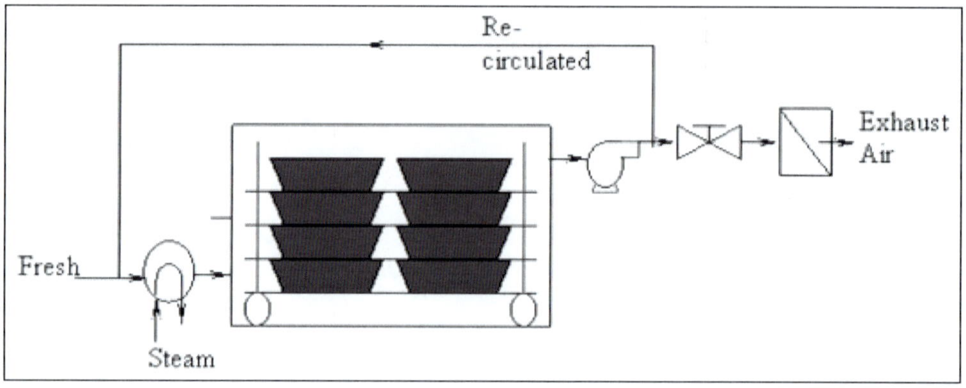

Tray Dryer

This type of dryer is frequently used for drying of wood and various agricultural products.

Band (Belt) Dryers

A band dryer is preferable if the particles to be dried are rather coarse (i.e. between 5 to 10 mm). The particles spread evenly into slowly moving, e.g. 5mm/s, perforated belt. The belt moves into a drying cabinet and warm gas passes downward through the layer. This type of dryer is chosen when it is not possible to suspend the particles in the drying gas. The dryer must offer a residence time, say 15min, because bound moisture must defuse through the pellet.

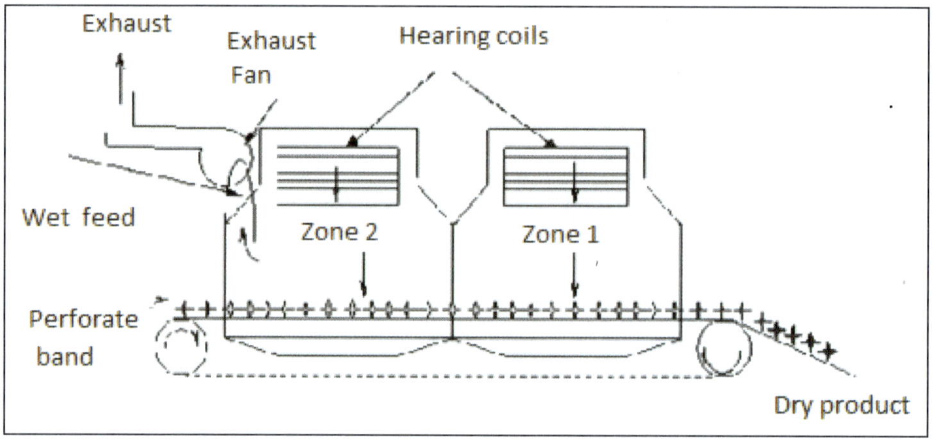

Band Dryer

Rotary Dryer

The rotary dryer consist of a cylindrical shell, horizontal or slightly inclined toward the outlet.

They are heated by:

a. Direct contact of air or gas with the solids,

b. By hot gasses passing through an external jacket on the shell, or,

c. By steam condensing in a set of longitudinal tubes mounted on the inner surface of the shell.

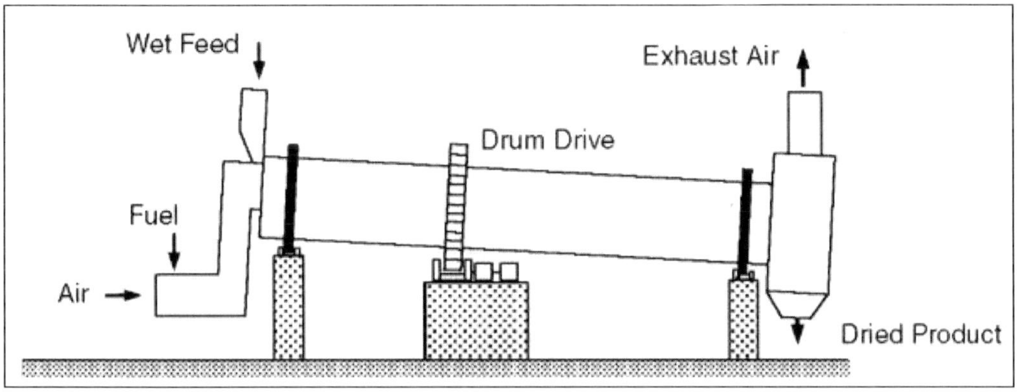

Rotary dryers

The product is fed to the upper orifice and transported during mixing through the cylinder. Regardless of the method of the heating, the water is removed with the air. The disadvantage of rotary dryers is the big power losses, which occur if the product is fine grained. The installation of the particle/gas separation at the moist air outlet (e.g. cyclones) may reduce the product losses significantly.

Roller Dryer

A roller dryer consist of a cylinder, heated from the inside by stream. A thin film of the product is sprayed on the outside of the cylinder and it is heated while the cylinder rotates. The rate of drying and the final water content in the product is affected by the:

a. Rotational spread of the roller,

b. Steam pressure,

c. Thickness of the film and

d. Properties of the product.

Owing to the sort contact time in combination with high drying temperature, the roller dryer is well adapted to heat sensitive products. Typical applications are fabric drying in textile industries and paper band drying in paper mills.

Hitesh N Panchal & Dr. P. K. Shah explained on the glass thickness lower glass cover thickness increases distillate output from solar still, i.e. 4 mm glass cover thickness produces more distillate output compared with 8 mm as well as 12 mm. Lower glass cover thickness decreases inner glass cover temperature inside solar still, i.e. 12 mm glass cover thickness produces highest inner glass cover temperature compared with 4 mm as well as 8 mm thickness of glass cover.

Fluidized Bed Dryer

In a fluidized –bed dryer, the particles are fluidized by air or gas in a boiling bed unit. The average time a particle stays in a bed is usually between 30 and 60s. If fine particles are present, there may be considerable solids carried over with the exit gas, and cyclones and bag filters are needed for their recover.

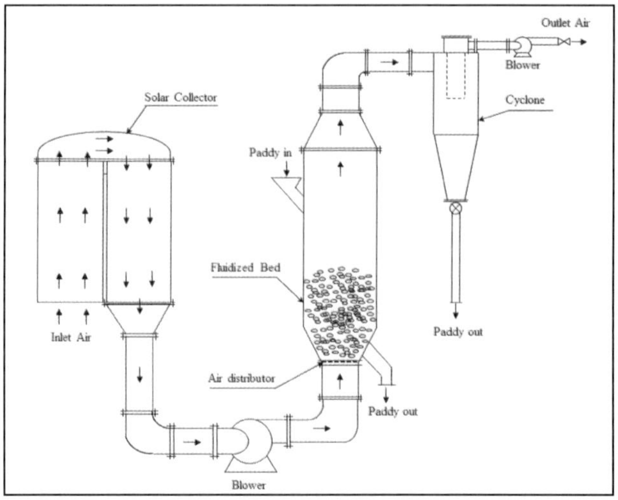

Spray Dryers

The pneumatic or 'flash' dryer is used with products that dry rapidly owing to the easy removal of free moisture or where any required diffusion to the surface occurs readily. Drying takes place in a matter of seconds. Wet material is mixed with a stream of heated air (or other gas), which conveys it through a drying duct where high heat and mass transfer rates rapidly dry the product. Applications include the drying of filter cakes, crystals, granules, pastes, sludges and slurries. In fact almost any material where a powdered product is required. Salient features are as follows:

1 Particulate matter can be dispersed, entrained and pneumatically conveyed in air. If this air is hot, material is dried.

2 Pre-forming or mixing with dried material may be needed feed the moist material.

3 The dried product is separated in a cyclone. This is followed by separation in further cyclones, fabric sleeve filters or wet scrubbers.

4 This is suitable for rapidly drying heat sensitive materials. Sticky, greasy material or that which may cause attrition (dust generation) is not suitable. The type of spray dryer is normally Pneumatic/Flash dryer is shown in figure below.

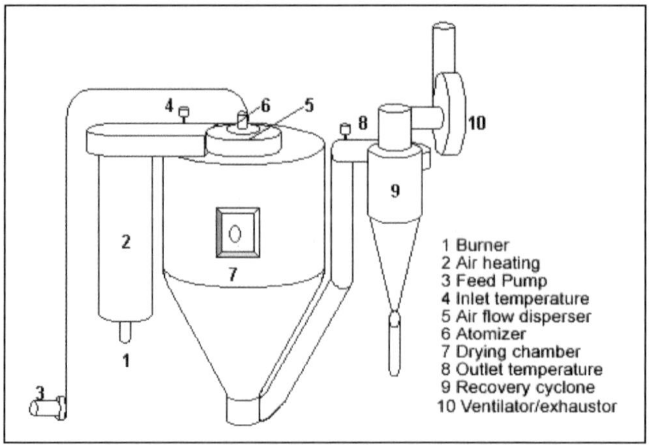

Spray Dryer

Electrical Methods

Dehydrator Dryer

A dehydrator dryer is dehydrating food is an ancient method of preserving food. It is a gentle, natural process which removes moisture from food. Using a controlled heat temperature, air is circulated from the top of the unit to each of the five trays and base. This method of drying seals in the flavor sand nutrients of the food, leaving a high food nutrient and vitamin content.

Natural healthy snacks can easily be created using your Food dehydrator. A variety of fruit rolls such as pear, berry and apple rolls, to name a few. You can make tasty, muesli bars, using all natural ingredients.

Dehydrator dryer

Oven Dryer

Oven drying is the simplest way to dry food because you need almost no special equipment. It is also faster than sun drying or using a food dryer. But oven drying can be used only on a small scale. An ordinary kitchen oven can hold only 4 to 6 pounds of food at one time. Set the oven on the lowest possible setting and preheat to 140 degrees F. (60 °C.). Do not use the broiler unit of an electric oven because the food on the top tray will dry too quickly' Remove the unit if it has no separate control. Some gas ovens have a pilot right, which may keep the oven warm enough to dry the food.

Oven Dryer

It is important to keep the oven temperature at 140 to 160 °F. (60 to 70 °C.). So put an oven thermometer on the top tray about half way back where you can see it easily. Check the temperature about every half hour. Arrange 1 to 2 pounds of prepared food in a single layer on each tray. Put one tray on each oven rack. Allow 1-1/2 inches of space on the sides, front, and back of the trays so that air can circulate all around them in the oven. To stack more trays in the oven, use blocks of wood in the comers of the racks to hold the trays at least 1 inches apart. Dry no more than four trays of food at a time. A lighter load dries faster than a full load. Keep the oven door open slightly during drying. A rolled newspaper, a block of wood, or a hot pad will keep the door ajar so that moist air can escape while the heat stays in the oven. Four to six inches for electric ovens or 1 to 2 inches for gas ovens is usually enough space for ventilation, but use a thermometer to check the oven temperature to make sure it stays at 140 °F. An electric fan placed in front of the oven door helps to keep the air circulating. Shifting the trays often is important for even drying because the temperature is not the same everywhere in the oven. Rotate the trays from top to bottom and from front to back every half hour. It helps to number the trays so you can keep track of the order in which you rotate them. Stirring fruit or vegetables every half hour or so also helps the food to dry evenly.

Spotlight Dryer

Manjula et al. proposed laboratory experiments contain the spotlight drying method shown in. The study shows that there is an inverse relationship between air temperature and drying time. The air temperatureas radiation intensity is an influential external parameter which is not the case of air velocity. Drying curves do not clearly indicate the effect of air velocity on drying time, and there is an inverse relationship between radiation intensity and drying time. Taking into account the evaporation process of water, a better understanding of drying phenomena was achieved by determining internal conductance C(X) and D(X).

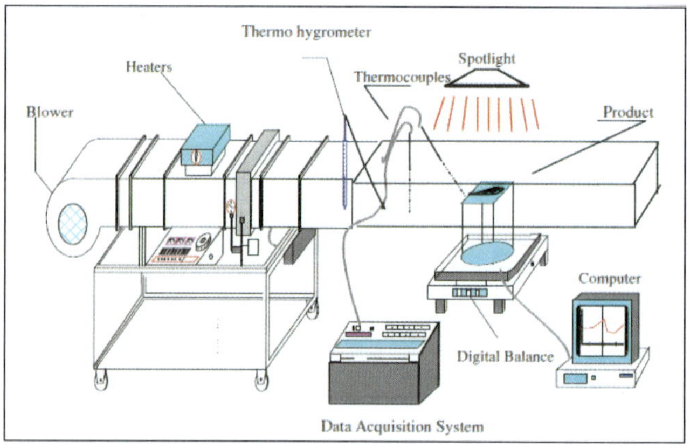

Spotlight dryer

No expression of C(X) and respectively D(X) with air velocity can be seen. Future work can be done designing and measuring drying experiments on a higher scale to better validate the results.

However, the laboratory model could satisfactorily represent the drying of red peppers in large ranges of air temperature, air velocity and incident radiation. The statistical analysis confirmed the suitability of the established model. The laboratory model does not roughly predict the drying data for open sun and greenhouse conditions because the drying operation presents inertia to variation of external parameters. To adjust predictions of the drying process under time varying conditions a correction factor in time was introduced in the formulation of the model. The corrected model could adequately describe the thin layer drying behavior of red peppers in open sun and greenhouse conditions. The methodology adopted in this particular drying process may be applied to different agricultural products as well. Models of this nature are applicable to the study and simulation of greenhouse type-solar dryers.

Natural Convection Methods

OYSD (Open Yard Sun Drying Method)

In OYSD, we are simply drying the products in open atmospheric space. Due to this the chances of products getting affected due to rain and dust are more. There is also a possibility fungus formation of the products if they are kept as it is.

Solar Tunnel Dryer

In natural convection method different types of solar dryers are used based on the quantity and time. Solar drying is a continuous process where moisture content, air and product temperature change simultaneously along with the two basic inputs to the system i.e. the solar isolation and the ambient temperature.

Solar Tunnel Dryer

The solar tunnel dryer consists of dryer comprises a plastic sheet-covered flat plate collector and a drying tunnel. The drying rate is affected by ambient climatic conditions. This includes: temperature, relative humidity, sunshine hours, available solar isolation, wind velocity, frequency and duration of rain showers during the drying period.

Open sun drying of various industrial and agricultural products is being practiced since age. Open sun drying is slow and exposes the produce to various losses and deterioration in quality. A number of industries have, therefore, accepted mechanical drying of the produce. Fuel wood, petroleum fuel, coal or electricity is used for air heating in the mechanical dryers. Solar air dryers have great potential for replacement of industrial scale drying of industrial and agricultural products. Besides, effecting saving of precious fossil fuel, fuel wood or electricity, the solar drying may also is cost effective.

Joy c.m. and K.P. Jose explained solar dryers in comparison with open sun drying gave better quality products with lesser drying time. Kamaruddin et al. have developed a method for the drying of pepper using solar energy. Pruthi had shown a drying time of 8 hours for 30-40 tones of pepper when dried in a mechanical dryer imported from Holland. To improve the overall quality of pepper, a solar dryer and some additional appropriate technologies were used to produce pepper of a high microbiological standard, deep black colour and low humidity. The quality of black pepper is assessed by its aroma and pungency retained after drying. The pungency of pepper is due to the presence of piperine.

Solar Tunnel Dryer

Oparaku, n.f explained about solar tunnel dryer model performance at 10% residual error interval was 78.4 and 83.3%, respectively, for global solar radiation and plenum chamber temperature. Linear relations existed between the simulated and mean of satellite global solar radiat ion, and simulated and actual plenum chamber temperature. The correlation between the simulated and satellite solar radiation was strong since the coefficient of determination was high $(R^2 = 0.788)$. S. A. S. M. Mohsin et al. explained about daily solar radiation varies between 4 and 6.5 kWh/m^2. In this regard, solar dryer for domestic as well industrial usage could be an effective alternative of saving conventional energy. Utilization of solar thermal energy through solar dryer is relatively in a nascent state in Bangladesh.

Mixed Mode Dryer

G. M. Kituu et al. discussed about mixed mode natural convection of solar dryer integrated with a simple biomass burner and bricks for storing heats. The dryer was designed for small-scale commercial producers of agricultural products in non-electrified locations. From a series of evaluation trials of the system, the capacity of the dryer was found to be 60– 65 kg of unshelled fresh harvested groundnuts. The drying efficiency of the solar component alone was found to be 23%. While, the efficiency of the burner with heat storage in producing useful heat for drying was found to be 40%. The key design features of the dryer contributed to produce an acceptable thermal efficiency, and uniformity of drying air temperature across the trays, were the jacket and gap enclosing the drying chamber and arranged bricks for storing heats.

S. Janjai et al. described the solar radiation passing through the polycarbonate roof heats the air and the products in the dryer as well as the concrete floor. Ambient air is drawn in through a small opening at the bottom of the front side of the dryer and is heated by the absorber and the products exposed to solar radiation. The heated air, while passing through and over the products absorbs moisture from the products. Direct exposure to solar radiation of the products and the heated drying air enhance the drying rate of the products.

A. A. Zomorodian and M. Dadashzadeh showed the moisture ratio decreases with a diminishing rate at different air flow rates and for two solar drying systems. This means that all drying periods were performed in the falling rate period. These results are in good agreement with the results of other researchers who had some extensive research on the solar drying of different products such as pistachio, chilli, crapes etc. The performance of the solar dryer depends on several factors, the solar radiation, inlet air temperature of solar dryer and the dryer design factors note that access to the highest temperature required for drying when tray 4 is from 61 °C and is at the highest intensity of the radiation 700 W/m2 at 11. There were no significant differences in temperature at the trays (1, 2 and 3) with the differences ranging from (2 to 5 °C).

B. K Bala et al proposed the solar tunnel drier was installed at the yard of the workshop of the Department of Farm Power and Machinery, Bangladesh Agricultural University, Mymensingh, Bangladesh. The drier was placed on raised platform. Mushrooms used for solar drying were collected from the local markets of Mymensingh and Savar. Three tests on solar drying of Mushrooms were carried out at Bangladesh Agricultural University, Mymensingh in the month of May in 2007 and in the month of March and May in 2008.

Mohod A.G et al showed during winter season, the average temperature inside solar tunnel dryer was found to be 50.39 °C with corresponding relative humidity 11.92 % which was the lowest value. The corresponding ambient temperature, relative humidity and solar intensity were found to be 33.97 °C, 14.38 %, 445.58 w/m^2 respectively. The maximum average temperature inside the solar tunnel dryer was found to be 52.41 °C at center of solar tunnel dryer followed by north side (51.06 °C) and south side (47.69 °C). The minimum average relative humidity inside the solar tunnel dryer was found to be 11.40 % at the center of solar tunnel dryer followed by north side (12.19 %) and south side (12.27 %). The increased relative humidity at south side could be attributed to incoming fresh air through air inlets provided at south side of solar tunnel dryer.

Ojike. O et al samples of pawpaw fruit were dried in the open-air and with solar dryers. Vitamin A, B1, B2, C and E were analysed to determine their concentrations before and after drying. In all cases there were significant changes in the concentration of vitamins after drying. The changes were much in open-air drying than in solar dryers used. Thus, the use of solar dryers for drying of pawpaw is highly recommended. Among the solar dryers used Latitudinal box dryer gave the best result in terms of vitamin retention.

Postharvest Technology

Postharvest technology is inter-disciplinary "Science and Technique" applied to agricultural produce after harvest for its protection, conservation, processing, packaging, distribution, marketing, and utilization to meet the food and nutritional requirements of the people in relation to their needs. It has to develop in consonance with the needs of each society to stimulate agricultural production; prevent Postharvest losses, improve nutrition and add value to the products. In this process, it must be able to generate employment, reduce poverty and stimulate growth of other related economic sectors. The process of developing of Postharvest technology and its purposeful use needs an inter disciplinary and multi-dimensional approach, which must include, scientific creativity, technological innovations, commercial entrepreneurship and institutions capable of in-ter-disciplinary research and development all of which must respond in an integrated manner to the developmental needs.

Importance

Importance of Postharvest technology lies in the fact that it has capability to meet food requirement of growing population by eliminating avoidable losses making more nutritive food items from low grade raw commodity by proper processing and fortification, diverting portion of food material being fed to cattle by way of processing and fortifying low grade food and organic wastes and by products into nutritive animal feed. Postharvest technology has potential to create rural industries. In India, where 80 percent of people live in the villages and 70 percent depend on agriculture have experienced that the process of industrialization has shifted the food, feed and fibre industries to urban areas. This process has resulted in capital drain from rural to urban areas, decreased employment opportunities in the rural areas, balance of trade in favour of urban sector and mismatched growth in economy and standard of living including the gap between rural and urban people. It is possible to evolve appropriate technologies, which can establish agricultural based rural industries.

The purpose of Postharvest processing is to maintain or enhance quality of the products and make it readily marketable. Prime example of postharvest processing of agricultural products is rice, a major crop in India. Paddy is harvested and processed into rice. Experiments with paddy crop in farmer's field in India have shown that if the crop is harvested at 20 to 22 per cent moisture as traditionally done, the field yield is increased by 10 to 20 percent. Similar is the case with respect to wheat, jowar and other crops.

Postharvest Losses

Due to old and outdated method of paddy milling, improper and inefficient methods of storage of paddy, rice, transport and handling we lose about nine percent of production. It is estimated that ten percent of food grains produced in India are lost in processing and storage. The traditional methods of storage are responsible for about six percent losses. If better methods of processing and storage are adopted, the losses could be reduced to 2 to 3 percent and more food grains could be available to the people. It is estimated that 10-15 percent of horticultural crop such as vegetables and fruits perish due to lack of proper methods of processing and storing. The loss in monetary term is estimated to be about Rs.20 crores annually.

Proper methods of processing, storage, packaging, transport and marketing are required for export of crops such as jute, tea, cashew nuts, tobacco, mango, litchi, nut, spices and condiments. One of the attributes to this postharvest system, as it is now constituted, is the large amount of wastage it involves. In case of food grains, some estimates suggest that in developing countries as much as 1/4th to 1/3rd of total crop may be lost as a result of inefficiencies in the postharvest system.

Losses of food crops refer to many different kinds of loss produced by a variety of factors. These include weight loss, loss of food values, loss of economic value, loss of quality or acceptability and actual loss of seeds themselves.

Priorities and Strategies

The priority areas in food processing are:

1. Processing of special fruits and nuts like, banana, litchi, mango, pineapple, makhana etc. and canning and storage facility for the above produce.

2. Large scale introduction of mini rice mill in villages and mandies coupled with semi-modern parboiling plant for paddy to have higher head rice recovery with better quality bran. Oil production from bran with a chain of collection mechanism for supplying raw material for the plant.
3. More emphasis on the use of power Ghani or expeller in place of Kolhu for higher recovery.
4. Establishment of dal mills in pulse growing belt as a village cooperative programme.
5. Emphasis on cottage industry involving village women for the manufacture of food products.
6. Popularization of low cost engineering storage structures.
7. Starch production from maize and potato and simultaneous oil production from maize.
8. Strengthening of research base with adequate financial support.
9. Emphasis on production of value added products from locally available fruits and vegetables.

There are many other areas of processing aspects which should be given priority. Once the processing, research and industry programme picks up many other outlets shall come up automatically. Processing industry has very good employment potential.

Postharvest Industries

1. The postharvest industry includes the following main components,
2. Harvesting and threshing,
3. Drying and storage,
4. Processing (conservation and transformation of the produce),
5. Utilization by consumer including home processing.

Other components of the system include:

1. Transportation and distribution,
2. Marketing,
3. Grading and quality control,
4. Pest control,
5. Packaging,
6. Communication among all concerned,
7. Information, demonstration and advisory systems,
8. Manufacture and supply of essential equipment and machinery,

9. Financial control,
10. Price stabilization,
11. Management and integration of the total system.

Potential of income and employment generation through postharvest operations-

Use of appropriate postharvest technology reduces the post harvest and storage losses; adds value to the product, generate employment in village and reestablishes agro-industries in rural sector. Presently, the farmers sell their products without processing. If they do primary processing and value addition in the villages, it will generate more income and employment in rural sector. The processing of food, feed, fibre, oilseeds and sugarcane will generate enough employment in rural areas. If an agro processing center is established in each big village or a cluster of small villages for primary processing, it will generate employment to about 4-5 persons and will increase income of the farmer/processor by about 15-20 percent. Use of proper postharvest technology of perishables and semi-perishables will reduce the wastage to great extent.

Potential of Income and Employment Generation in other Areas

Besides potential for income and employment generation in crop production and postharvest sector, there is a great potential of income and employment generation in allied sectors also by using Agricultural Engineering Technology. They are:

1. Animal production - Fodder, boiling, briquething, palleting, - Poultry feed Industry,
2. Fish Production - Improved hatchery, production and transport of fish seed fingerlings etc.,
3. Dairying - Processing of milk and making dairy products,
4. Energy management in agriculture - Efficient use of biomass, wind and solar energy,
5. Wasteland development - Conservation of wasteland, afforestation, management of trees and grasses,
6. Agro-forestry,
7. Watershed Management.

Action taken to tap the Potential- In order to take advantage of agricultural engineering technology for generating income and employment in rural areas following actions are suggested:

1. Bringing awareness amongst the rural people about the new developments in agricultural engineering technology in different fields.
2. Organizing training programmes for the farmers/agricultural laboures/entrepreneurs about the use of new technology.
3. Mass production of different types of agricultural machinery for farmers/entrepreneurs.
4. Starting agro processing centers in each village for primary processing of food grains, fruits and vegetables.

5. Providing institutional credit for the purchase of agrl. machinery and starting agro-processing centres.
6. Developing market network for purchase/supply of processed material from agro-processing centers.
7. Developing proper network and infrastructures for popularization of agricultural machinery for crop production, and setting up agro-processing centers.

Packaging Techniques for Fruits and Vegetables

India is a land of large varieties of fruits and vegetables due to its vast soil and climatic diversity. With 38 and 71 million tons of production of fruits and vegetables, India is the second largest producer of fruits and vegetables next to Brazil and China respectively. It is also a matter of concern that there is a wide gap between availability and the per capita nutritional requirement of fruits. The low availability of quality fruits & vegetables is mainly due to considerably high postharvest losses, poor transportation facilities, improper storage and low processing capacity coupled with the growing population. Around 20-30% losses take place during harvesting, grading, packaging, transportation and marketing of fruits.

The fruits of increased production of fruits and vegetables and other agricultural produce will be realized only when they reach the consumer in good condition and at a reasonable price. The existing postharvest loss of fruits and vegetables could be considerably reduced by adopting improved packaging, handling and efficient system of transport. Packaging of fruits and vegetables is undertaken primarily to assemble the produce in convenient units for marketing and distribution.

Requirements of Packaging

The package must stand up to long distance transportation, multiple handling, and the climate changes of different storage places, transport methods and market conditions. In designing fruit packages one should consider both the physiological characteristics of the fruit as well as the whole distribution network.

The package must be capable of:

1. Protecting the product from the transport hazards,
2. Preventing the microbial and insect damage,
3. Minimizing the physiological and biochemical changes and losses in weight.

The present packaging systems for fresh vegetables in our country is unsuitable and unscientific. The uses of traditional forms of packages like bamboo baskets are still prevalent. The other types of packages generally used are wooden boxes and gunnysacks. The use of corrugated fiberboard boxes is limited. The use of baskets besides being unhygienic also does not allow adequate aeration and convenience of easy handling and stocking. Considering the long term needs of eco-systems and to achieve an overall economy, other alternatives available like corrugated fibre board boxes, corrugated polypropylene board boxes, plastic trays / crates / wooden sacks, moulded pulp trays / thermoformed plastic trays and stretched film and shrink wrapping would have to be considered.

Modern packages for fresh fruits and vegetables are expected to meet a wide range of requirements, which may be summarized as follows:

- The packages must have sufficient mechanical strength to protect the contents during handling, transport and while sacked.
- The construction material must not contain chemicals, which would transfer to the produce and cause toxic to it or to humans.
- The package must meet handling and marketing in terms of weight, size and shape.
- The packages should allow rapid coding of the contents.
- The security of the package or its ease of opening and closing might be important in some marketing situations.
- The package should identify its contents.
- The package must be required to aid retail presentation.
- The package might need to be designed for ease of disposal, reuse or recycling.
- The cost of the package should be as less as possible.

Packaging may or may not delay or prevent fresh fruits and vegetables from spoiling. However, incorrect packaging will accelerate spoilage. Packaging should serve to protect against contamination, damage and excess moisture loss.

Packaging Materials in Use

A great variety of materials are used for the packing of perishable commodities. They include wood, bamboo, rigid and foam plastic, solid cardboard and corrugated fibre board. The kind of material or structure adopted depends on the method of perforation, the distance to its destination, the value of the product and the requirement of the market.

1. CFBC Boxes: Corrugated fiberboard is the most widely used material for fruit & vegetable packages because of the following characteristics:

- Light in weight,
- Reasonably strong,
- Flexibility of shape and size,
- Easy to store and use,
- Good pointing capability,
- Economical.

2. Wooden Boxes: Materials used for manufacture of wooden boxes include natural wood and industrially manufactured wood based sheet materials.

3. Sacks: Sacks are traditionally made of jute fibre or similar natural materials. Most jute sacks are provided in a plain weave. For one tonne transportation of vegetables, materials of 250 grams per square meter or less are used. Natural fibre sacks have in many cases been replaced by sacks made of synthetic materials and paper due to cost factors, appearance, mechanical properties and risk of infestation and spreading of insects. Sacks made of polypropylene of type plain weave are extensively used for root vegetables. The most common fabric weight is 70-80 grams per square meter.

Palletisation

Pallets are widely used for the transport of fruit & vegetable packages, in all developed countries.

The advantages of handling packages on pallets are:

- Labour cost in handling is greatly reduced.
- Transport cost is reduced.
- Goods are protected and damage reduced.
- Mechanized handling is very rapid.
- Through high stacking, storage space can be more efficiently used.
- Pallets encourage the introduction of standard package sizes.

Ventilation of Packages

Reduction of moisture loss from the product is a principal requirement of limited permeability packaging materials. A solution to moisture loss problems from produce appeared with the development and wide distribution of semi permeable plastic films. Airflow through the ventilation holes allows hot fruit or vegetable to slowly cool and avoid the buildup of heat produced by the commodity in respiration. Holes are also important in cooling the fruit when the packages are placed in a cold storage, especially with forced air-cooling. Ventilation holes improve the dispersal of ethylene produced.

Cushioning Materials

The function of cushioning materials is to fix the commodities inside the packages and prevent them from mixing about in relation to each other and the package itself, when there is a vibration or impact. Some cushioning materials can also provide packages with additional stacking strength. The cushioning materials used vary with the commodity and may be made of wrapping papers, fibreboard (single or double wall), moulded paper pulp trays, moulded foam polystyrene trays, moulded plastic trays, foam plastic sheet, plastic bubble pads, fine shredded wood, plastic film liners or bags.

Controlled and Modified Atmospheric Packaging (CAP and MAP)

The normal composition of air is 78% Nitrogen, 21% Oxygen, 0.03% Carbon dioxide and traces of other noble gases. Modified atmosphere packaging is the method for extending the shelf-life of perishable and semi-perishable food products by altering the relative proportions of atmospheric gases that surround the produce. Although the terms Controlled Atmosphere (CA) and Modified

Atmosphere (MA) are often used interchangeably a precise difference exists between these two terms. Controlled Atmosphere (CA).

This refers to a storage atmosphere that is different from the normal atmosphere in its composition, wherein the component gases are precisely adjusted to specific concentrations and maintained throughout the storage and distribution of the perishable foods. Controlled atmosphere relies on the continuous measurement of the composition of the storage atmosphere and injection of the appropriate gases or gas mixtures into it, if and when needed. Hence, the system requires sophisticated instruments to monitor the gas levels and is therefore practical only for refrigerated bulk storage or shipment of commodities in large containers.

If the composition of atmosphere in CA system is not closely controlled or if the storage atmosphere is accidentally modified, potential benefit can turn into actual disaster. The degree of susceptibility to injury and the specific symptoms vary, not only between cultivars, but even between growing areas for the same cultivars and between years for a given location. With tomatoes, excessively low O_2 or high CO_2 prevents proper ripening even after removal of the fruit to air, and CA enhances the danger of chilling injury.

Modified Atmospheric Packaging (MAP)

Unlike CAPs, there is no means to control precisely the atmospheric components at a specific concentration in MAP once a package has been hermetically sealed.

Modified atmosphere conditions are created inside the packages by the commodity itself and / or by active modification. Commodity generated or passive MA (Modified Atmosphere) is evolved as a consequence of the commodity's respiration. Active modification involves creating a slight vacuum inside the package and replacing it with a desired mixture of gases, so as to establish desired EMA (Equilibrated Modified Atmosphere) quickly composed to a passively generated EMA.

Another active modification technique is the use of carbon dioxide or ethyl absorbers (scavengers) within the package to prevent the build-up of the particular gas within the package. This method is called active packaging. Compounds like hydrated lime, activated charcoal, magnesium oxide are known to absorb carbon dioxide while iron powder is known as a scavenger to carbon dioxide. Potassium permanganate and phenyl methyl silicone can be used to absorb ethylene within the packages. These scavengers can be held in small sachets within the packages or impregnated in the wrappers or into porous materials like vermiculite. For the actively respiring commodities like fruits and vegetables, the package atmosphere should contain oxygen and carbon dioxide at levels optimum to the particular commodity. In general, MA containing between 2-5% oxygen and 3.8% carbon dioxide have shown to extend the shelf life of a wide variety of fruits and vegetables.

If the shelf life of a commodity at 20-25 °C is one day, then by employing MAP, it will get doubled, whereas refrigeration can extend the shelf life to 3, and refrigeration combined with MAP can increase it to four days. Few types of films are routinely used for MAP. The important ones are polyvinyl chloride, (PVC), polystyrene, (PS), polyethylene (PE) and polypropylene (PP). The recent developments in co-extrusion technology have made it possible to manufacture films with designed transmission rates of oxygen.

Vacuum packaging

Vacuum packaging offers an extensive barrier against corrosion, oxidation, moisture, drying out, dirt, attraction of dust by electric charge, ultra violet rays and mechanical damages, fungus growth or perishability etc. This technology has commendable relevance for tropical countries with high atmospheric humidity.

In vacuum packaging, the product to be packed is put in a vacuum bag (made of special, hermetic fills) that is then evacuated in a vacuum chamber and then sealed hermetically in order to provide a total barrier against air and moisture. If some of the product cannot bear the atmospheric pressure due to vacuum inside the package then the packages are flushed with inert gases like Nitrogen and CO_2 after evacuation.

Edible Packaging

An edible film or coating is simply defined as a thin continuous layer of edible material formed on, placed on, or between the foods or food components. The package is an integral part of the food, which can be eaten as a part of the whole food product. Selection of material for use in edible packaging is based on its properties to act as barrier to moisture and gases, mechanical strength, physical properties, and resistance to microbial growth. The types of materials used for edible packaging include lipids, proteins and polysaccharides or a combination of any two or all of these. Many lipid compounds, such as animal and vegetable fats, acetoglycerides have been used in the formulation of edible packaging for fresh produces because of their excellent moisture barrier properties. Lipid coatings on fresh fruits and vegetables reduce weight losses due to dehydration during storage by 40-70 per cent. Research and development effort is required to develop edible films and coatings that have good packaging performance besides being economical.

Improved packaging will become more essential in India as International trade expands after globalization. Standardized packaging of sized and graded produce that will protect the quality during marketing can greatly aid transactions between sellers and buyers. Better packaging should be of immediate value in reducing waste. Much background research on packaging of perishable products and flowers is needed simulating the actual handling conditions expected during marketing.

Value Addition

After harvest the biological produce can be either preserved or processed. Value addition is a terminology used to define the processing of biological produce. Through processing the value of the commodities can be increased by converting it to different products by using conventional or modern processing techniques, thereby the storage life of the produce is enhanced.

Value Added Products

1. Fruit Juice: It is a natural juice obtained by pressing out the fruits. Fruit juices may be sweetened or unsweetened.
2. RTS: It is prepared from fruit juices which must have atleast 10 per cent fruit juice and 10 per cent total sugar.

3. Fruit Juice Powder: The fruit juice is converted into highly hygroscopic powder. These are kept freeze dried and used for fruit juice drinks by reconstituting their composition.

4. Fermented fruit beverages: These are prepared by alcoholic fermentation by yeast of fruit juice. The product thus contains varying amounts of alcohols e.g. Grape wine, orange wine and berry wines from strawberry, blackberry etc.

5. Jam: Jam is a concentrated fruit pulp processing a fairly heavy body form rich in natural fruit flavour. It is prepared by boiling the fruit pulp with sufficient quantity of sugar to a reasonably thick consistency to hold tissues of fruit in position.

6. Jelly: Jelly is a semi solid product prepared by cooking clear fruit extract and sugar.

7. Marmalade: It is usually made from citrus fruits and consists of jelly containing shreds of peels suspended.

8. Tomato Ketchup: It is prepared from tomato juice or pulp without seeds or pieces of skin. Ketchup should contain not less than 12 per cent tomato solids and 28 per cent total solids.

9. Pickles: Food preserved in common salt or in vinegar is called pickle. Spices and oil may be added to the pickle.

References

- Design-and-development-of-potato-harvester-ijariie: ijariie.com, Retrieved 29 April, 2019
- Fruit-harvesting-robots: intorobotics.com, Retrieved 30 June, 2019
- Combine-harvester: newworldencyclopedia.org, Retrieved 19 August, 2019
- Threshing: farmech.gov.in, Retrieved 10 may, 2019
- Power-thresher-types-and-components-machines-agriculture, power-thresher, machinery: agricultureinindia.net, Retrieved 14 February, 2019

Permissions

All chapters in this book are published with permission under the Creative Commons Attribution Share Alike License or equivalent. Every chapter published in this book has been scrutinized by our experts. Their significance has been extensively debated. The topics covered herein carry significant information for a comprehensive understanding. They may even be implemented as practical applications or may be referred to as a beginning point for further studies.

We would like to thank the editorial team for lending their expertise to make the book truly unique. They have played a crucial role in the development of this book. Without their invaluable contributions this book wouldn't have been possible. They have made vital efforts to compile up to date information on the varied aspects of this subject to make this book a valuable addition to the collection of many professionals and students.

This book was conceptualized with the vision of imparting up-to-date and integrated information in this field. To ensure the same, a matchless editorial board was set up. Every individual on the board went through rigorous rounds of assessment to prove their worth. After which they invested a large part of their time researching and compiling the most relevant data for our readers.

The editorial board has been involved in producing this book since its inception. They have spent rigorous hours researching and exploring the diverse topics which have resulted in the successful publishing of this book. They have passed on their knowledge of decades through this book. To expedite this challenging task, the publisher supported the team at every step. A small team of assistant editors was also appointed to further simplify the editing procedure and attain best results for the readers.

Apart from the editorial board, the designing team has also invested a significant amount of their time in understanding the subject and creating the most relevant covers. They scrutinized every image to scout for the most suitable representation of the subject and create an appropriate cover for the book.

The publishing team has been an ardent support to the editorial, designing and production team. Their endless efforts to recruit the best for this project, has resulted in the accomplishment of this book. They are a veteran in the field of academics and their pool of knowledge is as vast as their experience in printing. Their expertise and guidance has proved useful at every step. Their uncompromising quality standards have made this book an exceptional effort. Their encouragement from time to time has been an inspiration for everyone.

The publisher and the editorial board hope that this book will prove to be a valuable piece of knowledge for students, practitioners and scholars across the globe.

Index